Abdelmounaim Khemmouli
Yamina Haider

Agriculture 3E with Camelina: Water, Energy, Hope in Arid Zones

Abdelmounaim Khemmouli
Yamina Haider

Agriculture 3E with Camelina: Water, Energy, Hope in Arid Zones

ScienciaScripts

Imprint
Any brand names and product names mentioned in this book are subject to trademark, brand or patent protection and are trademarks or registered trademarks of their respective holders. The use of brand names, product names, common names, trade names, product descriptions etc. even without a particular marking in this work is in no way to be construed to mean that such names may be regarded as unrestricted in respect of trademark and brand protection legislation and could thus be used by anyone.

Cover image: www.ingimage.com

This book is a translation from the original published under ISBN 978-620-8-82572-0.

Publisher:
Sciencia Scripts
is a trademark of
Dodo Books Indian Ocean Ltd. and OmniScriptum S.R.L publishing group

120 High Road, East Finchley, London, N2 9ED, United Kingdom
Str. Armeneasca 28/1, office 1, Chisinau MD-2012, Republic of Moldova, Europe
Managing Directors: Ieva Konstantinova, Victoria Ursu
info@omniscriptum.com

Printed at: see last page
ISBN: 978-620-8-99033-6

Summary:

In Algeria, food security is a major challenge for sustainable development, particularly in the face of exponential population growth. According to the United Nations, Algeria ranks first among African countries in terms of food security, but this independence remains fragile in the face of growing challenges linked to climate change and the variability of water resources. Severe climatic deficits and the impacts of climate change are leading to a fall in agricultural yields, justifying the adoption of innovative hydro-agricultural development practices. In this context, the introduction of new plant species adapted to semi-arid conditions is a promising way of optimising water use. Among these species, camelina *(Camelina sativa L. Crantz),* an oilseed crop renowned for its resistance to drought, was selected in this study to assess its response to a continuous deficit irrigation regime. The main objective is to modulate water inputs in order to identify optimal strategies for maintaining agronomic performance while reducing water consumption. The study was conducted in the Aurès region, a semi-arid environment marked by irregular rainfall well below the global annual average, as well as by extreme climatic events. The Alba genotype of camelina was studied over two consecutive agricultural seasons (2021/2022 and 2022/2023) as part of experiments carried out on the site of the Institute of Veterinary and Agronomic Sciences at the University of Batna 1. The experimental protocol is based on four distinct water regimes: (T0 = rainfed), (T1 = fully irrigated), (T2 = 75% ETc) and (T3 = 35% ETc). The experimental set-up, structured in completely randomised blocks with four replications, also incorporates three drainage lysimeters for precise measurement of Potential Evapotranspiration (PET). These methodological approaches make it possible to examine the agronomic and environmental behaviour of camelina under limited irrigation, highlighting its potential as a crop adapted to semi-arid zones. Camelina showed a significant response to limited irrigation, assessed through six main parameters: (1) morphological, fertility and productivity characteristics, (2) irrigation water use efficiency (IWUE), (3) nutritional quality (oil and protein content), (4) fodder quality of by-products (straw and cake), (5) energy consumption, and (6) energy potential (biofuel production).The results show that treatment T2 reduced grain yield and oil content by 20.77% and 5.28% respectively, on average over two agricultural seasons, compared with treatment T1 . On the other hand, T2 improved EUEI (+8.20%) and water productivity in feed units for meal (+21.96%), without affecting the energy value of the straw. In addition, T2 increased the protein content of the seeds (+6.75%) while reducing electricity consumption by more than 30% compared with T1. However, functional reliability to optimise the use of water and energy resources, making it possible to reconcile irrigation efficiency, sustainability and economic profitability in a context where there is an urgent need to strike a balance between water availability, agricultural yields and the search for reliable alternative solutions for animal feed in semi-arid regions. It is recommended that moderate irrigation restrictions (T2 = 75% ETc) be preferred to severe restrictions (T3 = 35% ETc). This approach strengthens the nexus between food security and agricultural water management.

Key words: *Camelina sativa*, deficit irrigation, irrigation water productivity in forage units, irrigation water use efficiency (IWUE), sustainable development.

Contents

General introduction

Algeria, located on the southern shore of the Mediterranean, faces growing challenges in maintaining a balanced agricultural trade balance, a problem shared with other countries in the region. These challenges have intensified over the years (Nadhira & Ouahiba, 2018) .

Algeria's increasing dependence on agri-food imports makes it vulnerable to fluctuations in international markets and unforeseen external crises, such as the global economic crisis and the COVID-19 pandemic.

The COVID-19 pandemic exacerbated this vulnerability, revealing the weakness of global supply chains, exacerbated by the conflict in Ukraine. Logistical disruption, border closures and price volatility have directly affected the availability of food products.

Food vulnerability in Algeria is not limited to cereals alone; it is a complex phenomenon that also extends to the oilseed sector and its co-products. Indeed, Algeria is a net importer of oilseeds, whether in the form of crude oil or soy, sunflower and rapeseed cakes, reflecting a significant deficit in edible oil and fodder for livestock in the national economy. Despite the efforts made, olive oil production remains insufficient to make up this shortfall, even with the acceleration in planting in recent years.

The cost of importing oilseeds and their derivatives reached 1.4 billion dollars in 2011, ranking third after cereals and dairy products, with 61% for crude oils and 36% for oilcakes (a sixfold multiplication factor between 2000 and 2011) (Rastoin & Benabderrazik, 2014) . According to FAO, (2022) , the prices of staple foods around the world have risen sharply. The biggest increase has been for oilseeds and their co-products, where prices have risen by 125.9% over the last three years.

Algeria is committed to the systematic regulation of prices and profit margins for refined oils. Margins are capped at 8% of cost price, in accordance with the decree of 7 November 1996. Compensation is granted to manufacturers to absorb price volatility on the international market and maintain the price of a litre of edible oil at DZD 120/litre, in accordance with Decree 11-108.

This is a heavy burden on the public purse, exacerbated by the fact that crude oils and oilcakes are imported separately, rather than oilseeds being imported and crushed locally. This results in a double loss of foreign currency on the one hand, and a loss of local crushing units for local economic development on the other.

As a result, the Ministry of Agriculture and Rural Development has adopted a roadmap for the five-year period (2020-2024), including recommendations to introduce oilseed crops in order to reduce the heavy import bill and produce 30% of national needs by 2024.

Against this backdrop, it is crucial to steer oil and fodder production towards new prospects, in particular by introducing new species adapted to Algeria's different ecological zones. The aim is to diversify food and fodder resources, while looking for alternatives to make up for the shortfall, which seems to be one of the priorities of the agricultural sector in semi-arid environments.

Indeed, before 2021, oilseed crops will be practically non-existent in the Algerian agricultural system, and their introduction will be particularly difficult due to the predominance of rain-fed agriculture in local farming and a climate that is tending to become increasingly arid. McDermid et *al.*, (2023) have pointed out that crop water requirements will increase by 5% by 2050 to maintain current levels of productivity, exacerbated by alarming trends, including a reduction in rainfall of 10% to 30% by 2080 (UNEPMAP - UN, n.d.) . This challenge is becoming even more difficult in the face of the combined effects of climate change, population growth and competition from other economic sectors such as industry and urban development.

The situation is worsening, with irrigation becoming essential to compensate for rainfall deficits and boost the resilience of the agricultural system in semi-arid environments, while adapting to areas vulnerable to climate change. This approach makes a crucial contribution to the integrated management of the food security-water management nexus. A holistic approach involves supporting sustainable agricultural practices and water-efficient strategies, ensuring equitable access to water and food while preserving natural resources for future generations.

Faced with these multifactorial problems, scientific research in agronomy must aim to develop an agriculture capable of producing optimum yields even under difficult development conditions.

A second green revolution focusing on nutrient and water use efficiency is proposed as one of the solutions to meeting growing food demand and combating food shortages in the coming decades (Blum, 2009) .

Lopez et *al.*, (2017) found that current rates of agricultural water use are unsustainable in many countries, including Algeria, highlighting the urgent need to improve water management strategies for greater efficiency. Lopez-Gunn et *al*., (2012) state that it is essential to implement more effective strategies that allow sufficient food production while preserving water resources. This approach has become crucial since the 2000s, when water management in agriculture began to focus on increasing water productivity.

It is crucial that irrigation management in semi-arid environments moves from a focus on production per unit area to maximising production per unit of water consumed, known as water productivity (Rekaby et al., 2016) . Among the effective practices for maximising irrigation water productivity, deficit irrigation stands out in particular.

Fernandes-Silva et *al*., (2018) describe sustained deficit irrigation as an irrigation strategy consisting of using a reduced volume of water, controlled by indicators of water stress or according to a percentage of total water requirements. This method accepts a certain level of yield reduction, but also saves a proportion of irrigation water and a fraction of electrical energy. The water and energy saved can be used to irrigate new areas, thereby improving crop diversification, offsetting the loss of productivity and reducing the financial cost of pumping, as well as the environmental impact and depletion of gas reserves. Deficit irrigation is a powerful tool for strengthening the water-food security nexus. It will be even more effective when combined with the introduction of new, less water-hungry species. This approach supports and reinforces the sustainable use of water while improving the resilience of agricultural systems in the face of climate change.

In this context, among the most promising species of recent years, the versatile *Camelina sativa* from the Brassicaceae family is emerging as a reliable alternative to conventional oilseeds in semi-arid regions. Camelina, a former crop that disappeared in favour of more productive species such as rapeseed, has attracted growing interest in recent years thanks to its wide suitability for a variety of climatic and soil conditions (C. Zhang & Auer, 2020) . Highly resistant to diseases and pests, camelina is the least sensitive of all cruciferous species to water shortages. It is attracting the attention of the scientific community, as shown by the significant increase in publications devoted to this plant over the last decade (Sydor et al., 2022) . Because of its potential to thrive on marginal lands with limited water resources (Nithyapriya et al., 2021) , with an approximate requirement of 332-490 mm (French et al., 2009; D. J. Hunsaker et al., 2013) , camelina is proving to be a valuable option for agriculture in semi-arid environments.

Despite this growing popularity, camelina remains little-known in Algeria, which represents an opportunity to meet the major challenges not only of agriculture, but also thanks to its wide range of food and industrial uses. Its high oil content, very rich in omega-3 and omega-6 fatty acids, contributes to the agri-food industry (C. Zhang & Auer, 2020) , pharmaceuticals (Obour et al., 2017) , cosmetics (Delmas et al., 2013) and bioenergy through its ability to produce an environmentally friendly biofuel (Agarwal et al., 2021) .

In addition to its agronomic and industrial advantages, camelina stands out for its wide range of uses in animal feed, particularly for fish farming, poultry farming and cattle and sheep rearing, either as grain or as oilcake (Singh et al., 2023) . These qualities could contribute effectively to the resilience of the Algerian agricultural system and to the diversification of fodder resources in Algeria.

Our study aimed to assess the divergences and convergences in the response of camelina yield quality and water productivity to limited water supplies, as well as the behaviour of the fodder quality of its by-products (straw and oilcake). In addition, the assessment of water productivity in forage units over two successive cropping seasons provided an understanding of how irrigation practices influence the general aspects of crop productivity and water use efficiency in the Aurès region in a semi-arid environment.

In order to provide producers and decision-makers with valuable information on the economic and productive consequences of water allocations in a context of continuously declining water resources in Algeria, our study also explores deficit irrigation strategies to optimise water use and strengthen resilience in the face of climate change. The results of this research are intended to provide practical recommendations for improving food security and agricultural sustainability in regions facing declining water resources.

Chapter I. Bibliographical summary

1. Food security and sustainable development:

Food security eand sustainable developpement are of paramount importance in the agricultural context, particularly in semi-arid regions where the rarity of ressources eand vulnerability to climatic variations peuvent have a significant impact on agricole production. In this context, Algeria, like many other countries on the southern Mediterranean rim, faces the complex challenge of feeding a growing population while preserving the long-term sustainability of its agricole system. This challenge is intensified by changing climate patterns and extreme events, which increase the vulnerability of the agricultural sector.

With this in mind, Algeria's agricultural policies must evolve in order not only to ensure food security, but also to strengthen the overall resilience of the agricultural system. Promote sustainable agricultural practices that make judicious use of available resources while limiting environmental impact, in order to meet current needs while guaranteeing those of future generations.

Increasing dependence on food imports exposes Algeria to ae seriese of majeur challenges, particularly en view of fluctuations in international markets eand the risques associated with externeal shocks such as the Ukrainian crises eand the Covid-19 pandemic. Ces facteurs, susceptibles of engendrer des abrupt increaseses in import costs, mette sereusementy testing price stability and food supplies, comme le montre clairement the table representing the FAO price index:

Table 1. Trends in food prices according to the FAO index

Products	Price rises in (%)	Price increase in FAO index (Points)
Vegetable oil	125.9	+144.0
Cereals	74.1	+73.2
Meat	45.0	+45.2
Dairy products	52.5	+53.3

(FAO, 2022)

The biggest price increase was for vegetable oils, which rose by 125.9%. This rise was due to a number of factors, including the war in Ukraine, which disrupted production and exports of sunflower oil, one of the main vegetable oils produced and exported by Ukraine.

Vegetable oils are the lipid component of oilseed crops, including seeds, nuts, oil palm fruit, olives and soya (FAO, 2022). They play an essential role in human and animal nutrition, providing lipids, vitamins and minerals, and are even used, albeit modestly, in the production of biodiesel. The growing importance of this sector has led to significant expansion over the last two decades (Figure 1). Vegetable oil production varies from region to region, reflecting the particular dietary habits of each area. (Figure 2).

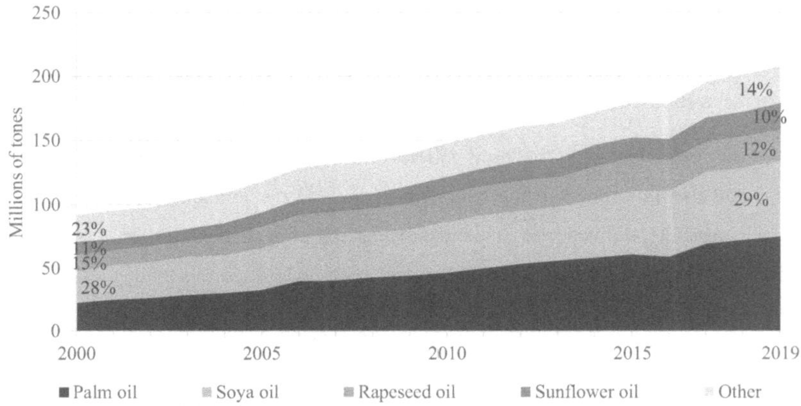

Figure 1: World production of vegetable oils, main products

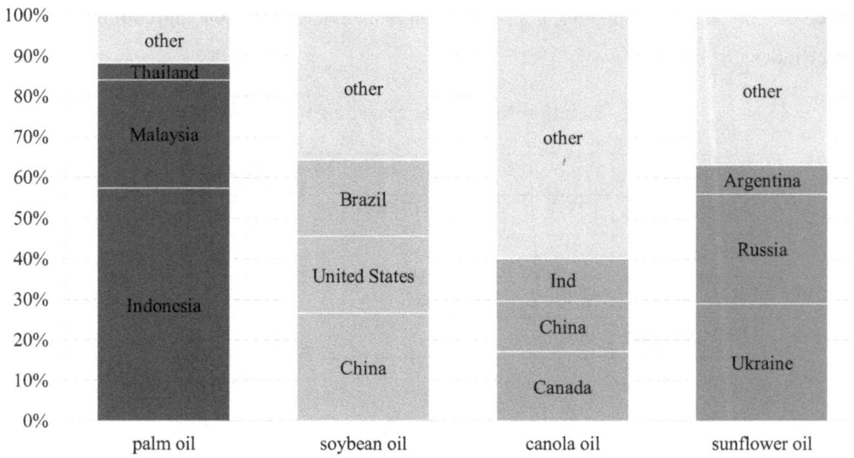

Figure 2: World production of the main vegetable oils by main producers (2019) (FAO, 2022)

There is a wide variety of vegetable oils in the world, extracted from different species:

Palm oil: produced in Indonesia and Malaysia.

Soya oil: Mainly produced in the United States, Brazil, Argentina and China.

Canola oil: the main producers are Canada, the European Union, India and China.

Sunflower oil: Russia and Ukraine dominate world production, along with Argentina.

Olive oil: a typical product of the Mediterranean region (Spain, Greece and Tunisia...).

However, it is important to note that production can vary from region to region depending on the influence of various factors including climatic conditions, trade policies and cooking oil trends in each region.

2. Vegetable oils in Algeria

With virtually no oilseed crops grown in Algeria before 2021, the country found itself almost totally dependent on imports to meet its domestic vegetable oil needs, leading to a gradual and significant increase in import costs . This situation has recently led the government to launch initiatives to promote the cultivation of rapeseed and sunflower, with the aim of boosting food security. Although olive oil accounts for 13% of consumption, it is surpassed by soya oil (49%), sunflower (13%), rapeseed (7%) and groundnut (5%), 90% of which is imported, with domestic demand estimated at 430,000 tonnes. (Rastoin & Benabderrazik, 2014) .

The aim of the oilseed industry is to press the seeds to extract oil, at the same time producing oilcake for animal feed (primary processing). The crude oils are then refined to make them suitable for consumption (secondary processing) (Figure 3).

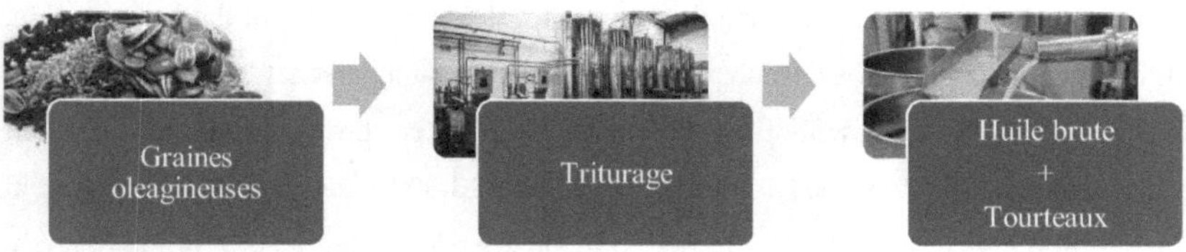

Figure 3. Oilseed crushing process

The absence of operational industrial oilseed crushing centres, and imports in the form of crude oils which are then refined locally, leads to a loss of the valuable co-product, oilcake. These then have to be imported separately to meet the feed requirements of livestock.

Faced with this situation, where the lack of local infrastructure for crushing oilseeds and the import of crude oils are leading to a loss of valuable co-products such as oilcake, the Ministry of Agriculture and Rural Development has drawn up a sectoral roadmap (2020-2024). This proposes recommendations for integrating oilseed crops into local farming practices, in order to reduce dependence on imports and cover 30% of national needs by 2024.

With this in mind, the forthcoming commissioning of oilseed crushing units in Jijel and Béjaïa by the Cevital group, scheduled for completion by the end of 2024, and the development of local production, represent a strategic opportunity to establish a genuine national oilseed industry. This project could reduce Algeria's food dependency, support rural employment, and be based on an inter-professional organisation integrating research and training in seeds, agronomy and processing technologies adapted to small units.

To achieve these objectives, the introduction of oilseed crops adapted to the Algerian context, such as canola, safflower, sunflower and groundnuts, is envisaged depending on regional conditions. However, the introduction of these crops faces major challenges linked to the preponderance of rain-fed agriculture and an increasingly arid climate, with average annual rainfall of just 89 mm, well below the global average (FAO, 2015) , and a nine-month dry period accompanied by high rainfall variability (Chennafi et *al* .., 2006)

3. Review of the agro-climatic situation in the Aurès region

The Mediterranean region is increasingly vulnerable to climate change, exacerbated by alarming trends, including a projected rise of 1.54°C by 2040, and a reduction in rainfall of 10% to 30% by 2080 (Gan et al., 2013; UNEPMAP - UN, n.d.) . with The overexploitation of water tables and the waste of water that has been rampant for many years have had the disastrous effect of causing a worrying drop in the level of water tables in the Batna region, which is experiencing a serious problem because pumping on a personal basis is carried out anarchically through 4,000 authorised boreholes, 300 water towers, thousands of illicit boreholes, some of which exceed 1,500 metres, at least 13.000 traditional wells operated in rural areas, 400 collective wells and 13,000 individual wells located mainly in urban areas. (DRE Batna, 2023), the scarcity of this resource is becoming increasingly apparent, and poses serious challenges for water optimisation, in particular its management has become a major issue in ensuring food security in the semi-arid environment.

The current situation has become increasingly critical, with irrigation becoming essential to compensate for rainfall deficits. This is necessary to meet the growing demand for food products estimated by the FAO, which is set to double over the next 40 years in developing countries. This means that farmers will have to produce twice as much from the same area in order to meet their food needs. (FAO, 2011) .

In Algeria, the pressure on water resources means that water management strategies need to be optimised to ensure the sustainability of agriculture. This involves improving the efficiency of agricultural water use, which is essential in regions with low rainfall, and practices aimed at improving water productivity, which has become a crucial issue in a context of limited availability.

The concept of sustainable water management refers to all practices that improve crop yields and minimise non-beneficial water losses (Johnson et al., 2001) . Thus improving irrigation water use efficiency, i.e. the ratio of water applied to crop yield (Stanhill, 1986) , and water productivity, i.e. the ratio of water consumed to crop yield (Kumar et al., 2010) , are both important parameters for achieving water resource sustainability.

In semi-arid zones, agriculture is dominated by rainfed systems, which depend entirely on rainwater stored in the soil. These areas are characterised by the spatio-temporal irregularity of rainfall in terms of quantity, with 38% of rainfall recorded when crops are planted and 24.6% of rainfall recorded during the winter, meaning that 63% of rainfall falls during a period when vegetation is not active due to the lower temperatures recorded. (Chennafi et *al* .., 2006)

. This unsuitable temporal distribution of rainfall makes irrigation essential to make up for the water deficit and meet the growing demand for food in these regions. However, optimised management of irrigation water, adjusted to the real needs of crops, remains difficult to achieve, limiting resource efficiency and water productivity...

Faced with these multifactorial problems, scientific research in agronomy must aim to develop a form of agriculture that achieves higher yields under difficult development conditions for the plant, by conserving all types of natural resources - water, soil and plant resources - in the context of sustainable and conservation agriculture.

Irrigation management in these areas needs to shift from a focus on production per unit area to maximising production per unit of water consumed, known as water productivity (Rekaby et al., 2016) . Among the effective practices for maximising water productivity is deficit irrigation.

4. The concept of deficit irrigation

The application of deficit irrigation can be very useful for reducing water demand in the agricultural sector. Deficit irrigation is the application of water below the full water requirements of crops at selected times (Forey et *al.*2016), another terminology Musick et *al.* (1994) deduced that deficit irrigation is the application of less water than necessary for and maximum yield, resulting in the conservation of a limited amount of irrigation water.

While Chai et *al.*(2016) have developed the concept of deficit irrigation and define it as the irrigation practice characterised by the application of irrigation water below the total amounts required for optimum growth and yield, aiming to positively improve plant response to some degree of water deficit, and improve the water use efficiency per crop. Adopting deficit irrigation principles means accepting a certain reduction in yield (Hamdy et al., 2005) . On the other hand, this reduction saves part of the irrigation water, which can be used to irrigate new areas, thereby increasing the cultivated area and compensating for the loss of productivity.

Plant-soil-water relationships need to be well understood to enable efficient use of water and effective management of agriculture. For this reason, deficit irrigation approaches have been developed to respond to possible changes in water resources. There is an ongoing effort to understand the relationships between soil water balance, crop yield and water use efficiency in order to develop better semi-arid crops and water management practices (Musick et al., 1994).

There are two main deficit irrigation strategies:

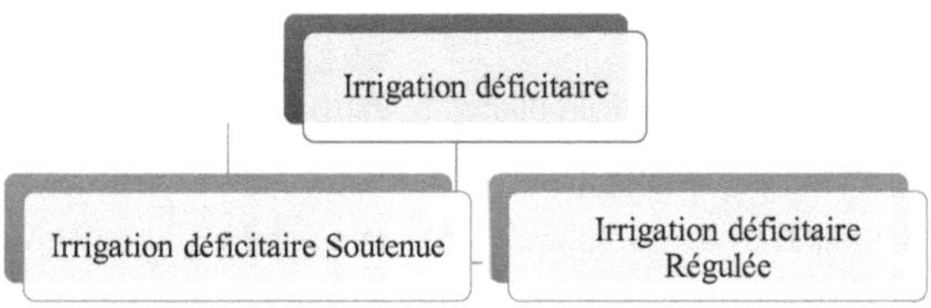

Figure 4. The main deficit irrigation strategies

4.1 Sustained deficit irrigation .

Fernandes-Silva et al.(2018) define sustained deficit irrigation as a strategy that involves the controlled distribution of a reduced volume of water, adjusted according to a water stress indicator or as a percentage of the crop's total water requirements throughout the irrigation season. The aim is to even out the water deficit over the entire crop cycle in order to avoid the appearance of severe water stress at a given time, which could have harmful consequences. Sofo et *al.* (2012) described sustained deficit irrigation as the practice of supplying a reduced volume of water, expressed as a percentage of crop evapotranspiration, throughout the irrigation season. Several studies have examined the effects of this method on crops.

In a study by Ramadan et *al.* (2014) , applying 75% of the water requirement of wheat under sprinkler irrigation in sandy soil resulted in only a 3% reduction in wheat yield. Similarly, Taha et *al.* (2015) found that an 11% reduction in water applied during deficit irrigation of wheat resulted in only a 2% yield loss.An experiment conducted by Noreldin et *al.* (2015) tested the effects of reducing irrigation water applied to wheat by 25% and 20% under drip and sprinkler systems, respectively. They observed a reduction in wheat yield of 20% and 18%, respectively. Other studies have compared irrigation regimes with different water supply restitutions with the fully irrigated regime for olive trees. Berenguer et *al.* (2006) concluded that a restitution of between 66% and 75% of crop evapotranspiration was sufficient to obtain yields similar to those of fully irrigated olive trees. Reported a significant reduction in phenolic compounds in oils under the highest levels of irrigation.

4.2 Irrigation deficit regulated.

Regulated deficit irrigation was first proposed by Chalmers et *al.*(1981) . This irrigation strategy reduces water inputs during specific periods when plants are less sensitive to water stress, with minimal effects on yield (Capra et al., 2008) . By applying sufficient quantities of water during the period when plants are sensitive to water deficit and by applying water deficits during periods when plants are more resistant to water deficiencies (Blum, 2009; Geerts & Raes, 2009) . Production parameters other than biomass may perform better with water restriction applied at certain stages, as pointed out by Tari.(2016) that water deficits applied at the bolting and heading stages of wheat increase the protein ratio and gluten content despite the significant reduction in grain yield and biomass as a result of this water restriction.Regulated deficit irrigation has been successfully adopted for certain crops to determine the stage of development that is most sensitive to stress, enabling irrigation to be carried out correctly to avoid a drop in yield by applying the minimum amount of water at the key time, successfully applied to various crops, such as barley (Kessai et *al.*, 2018) ,durum wheat (Chennafi et *al.*, 2006) ,Arbustives (Girona et *al.*, 2005) , sugar beet (Fabeiro et *al.*, 2003) , tomato (Mukherjee et *al.*, 2023) and cotton (Guedes et *al* .., 2023)

4.3. SWOT analysis of deficit irrigation .

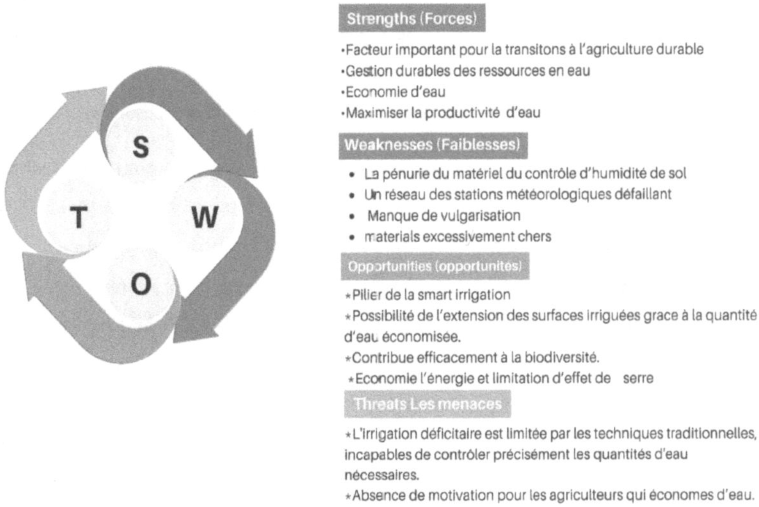

Figure 5. SWOT analysis of deficit irrigation in Algeria

5. Camelina: A Sustainable Approach for Semi-Arid Regions

Confronting the combined effects of climate change, population growth and competition from other economic sectors, such as industry and urban development, on water resources is an almost impossible challenge. In this complex configuration, the introduction of species that consume less water appears to be a solution to this multifactorial challenge, particularly in the face of a rapidly changing climate. However, this approach can be even more effective when combined with water-saving strategies that help to limit unsustainable water use while improving water resilience and promoting adaptation to climate change. In this context, the oilseed species camelina is emerging as a promising alternative in semi-arid regions, adapted to extreme environmental conditions.

Camelina is an annual herbaceous crop with winter and spring biotypes, each with different genotypes (Masella et *al* .., 2014) The life cycle varies from 90 to 250 days (Royo-Esnal & Valencia-Gredilla, 2018; Zanetti, Alberghini, Jérômela, et *al*., 2021) and in growing degree days (GDD), the average is 1200 GDD depending on the sowing season and the varieties adopted. The height that can be reached is between 0.65 and 1.05 m (Jankowski, 2019; Martinelli & Galasso, 2011) . During the flowering period, the flowers take on a pale yellow appearance with four petals (Figure 6A) that will reproduce into pods called siliques measured (4-5 mm) and numbering 60-115 siliques per plant (Figure 6B), which envelop 10-20 small camelina seeds (5 times smaller than peanut seeds) (Figure 6C) (Martinelli & Galasso, 2011; Załuski et *al.*, 2020; Zanetti et *al.*, 2017) with a thousand seed weight barely exceeding 1g (Czarnik et *al.*, 2018; Jankowski, 2019; Righini et *al.*, 2019; Załuski et *al.*, 2020; Zanetti et *al.*, 2017) , reflecting its low productivity compared to canola of between 13 and 33 qx/ha (Krzyżaniak et *al.*, 2019; Royo-Esnal & Valencia-Gredilla, 2018; Załuski et *al.*, 2020; Zubr, 1997) . Despite this low productivity, camelina seed develops a significant extraction rate of between 36 and 43% of the oil three times higher than the olive oil content (Kirkhus et *al.*, 2013; Krzyżaniak et *al.*, 2019; Załuski et *al.*, 2020; Zanetti et *al.*, 2017; Zubr, 1997) , widely used in food due to its richness in omega-3 and omega-6, suitable for salad dressing, cooking, baking and frying (Zubr, 2010) , and non-food in cosmetics (Hélène, 2020; Zubr, 1997) . Oilcake is also an important source of protein for animal feed (Singh et *al.*, 2023)).

Figure 6. Development phase of camelina

Camelina oil is not limited to food uses such as soap manufacture, cosmetics and paint production (Zubr, 1997) . The most interesting non-food use is as biodiesel, with properties complying with American Society for Testing and Materials (ASTM) standards (Martinelli & Galasso, 2011).

5.1. *Camelina sativa*: a versatile crop with multiple benefits

5.1.1. Agricultural

Camelina (Camelina sativa), a hardy plante, embodies ae precieuse ressource agricole thanks to ses multiples bienfaits eand its adaptability to des difficiles growing conditions. Cette culture nécessite peu of ewater et de pesticides (D. J. Hunsaker et *al.*, 2011) , evene on des marginal soils (C. Zhang & Auer, 2020) , thus offering ae durable et respectueuse option de the environnement. Its ability to prosper in des environnements that are arides et semi-arides en makes it ae precieuse allye for preservation des ressources naturelles.

The shorte periode de croissance de Caméline makes it rend ideale for rotational des cultures (Royo-Esnal & Valencia-Gredilla, 2018) , contributing to diversifier les pratiques agricoles et and preserver soil health. As ae covererture crop, elle playse a essentiel rôle in reducing erosion, renforcing the sustainability ofes agricoles pratiques, notamment in les arides et semi-arides regions confrontedes with des hydriques contraintes,

The 'allélopathie de Caméline, characterisede by the production of chemically allelopathice compounds, présente une novatrice approache for the gestion of mauvaises herbes en agriculture. Its use peut contribuer to reducinge dependence on chemic herbicides, thus favouring des more durables agricoles practices (Carton et *al* .., 2020)

What's more, Cameline's resistance to disease and pests en makes it ae robuste crop, requiring less phytosanitary treatment. Cela en fait une espèce attrayante pour le agriculture biologique, renforçant ainsi son rôle dans le contexte agricole.

5.1.2. Environment

Camelina (*Camelina sativa*) is a sustainable, environmentally-friendly crop offering a host of ecological benefits. The opportunity to transform the oil from this plant into a sustainable biofuel offers an environmentally friendly alternative, with lower greenhouse gas emissions than biodiesel derived from soya or canola (Krohn & Fripp, 2012) , thereby helping to combat climate change.

The by-products of oil extraction, such as oilcake, find significant use as animal feed, promoting a more efficient use of resources. The crop's low carbon footprint, combined with its minimal need for inputs such as fertilisers and pesticides, reinforces its environmentally-friendly profile (Righini et *al* .., 2019)

Camelina also stands out for its versatility, offering a range of opportunities for the development of environmentally-friendly products. With a reduced carbon footprint. In addition, Camelina's allelopathy offers an ecological strategy by limiting the negative impacts of herbicides on the environment and human health , while helping to counter the growing resistance of weeds to conventional herbicides (Uremis et *al* .., 2009)

5.1.3. Food

Camelina oil (*Camelina sativa*) has an exceptional nutritional composition, making it an ingredient of choice in a variety of food applications, as well as offering significant health benefits. With its high content of omega-3 polyunsaturated fatty acids and vitamins, Camelina oil has remarkable oxidative stability, making it an excellent ingredient for industry.

Alpha-linolenic acid (ALA), an essential member of the omega-3 family found in abundance in Camelina oil, plays a crucial role in brain function, heart health and reducing inflammation (Guendouz et *al.*, 2022) . The low erucic acid content is an advantage, avoiding the potential risks of heart disorders and lipidosis associated with specific fatty acids (Vollmann & Eynck, 2015) .

Its balanced fatty acid composition, with a low saturated fatty acid content compared to other vegetable oils, makes it a healthier choice for cooking and the production of processed foods. Camelina residues, such as oilcake, are of valuable use as animal feed, helping to reduce dependence on imported animal protein and boosting countries' food independence.

Adding Camelina to the human diet helps to diversify sources of nutrients, promoting a balanced diet. Its high content of tocopherols, including vitamin E, gives the oil greater stability against oxidation than other omega-3 rich oils (Hrastar et *al.*, 2009) . Studies also highlight the positive health effects of Camelina oil, including lowering total cholesterol levels in blood serum and reducing LDL (Low-Density Lipoprotein) in individuals with mild to moderate hypercholesterolaemia (Dobrzyńska & Przysławski, 2020) . These results reinforce the idea that incorporating Camelina oil into the diet can make a significant contribution to improving overall health.

5.1.4 Economic

The growing interest in sustainable products, with increased demand for *Camelina sativa*'s omega-3-rich oil, is opening up promising prospects. The growing demand for this crop in various industries offers an opportunity to diversify income and develop a new economic sector. *Camelina sativa* is an ideal raw material for the production of bio-based products such as biodegradable plastics and bio-based lubricants, thereby contributing to the transition towards a circular and sustainable economy.

Growing demand for sustainable, environmentally-friendly products is creating new opportunities for farmers and companies involved in growing *Camelina sativa*. This crop requires less water and chemicals than other conventional crops, reducing production costs. Its ability to grow in difficult and marginal conditions makes it an attractive crop for arid and poorly fertile regions, providing economic opportunities for these areas.

5.2 *Camelina sativa* in a cropping system

Camelina is a versatile plant that offers several possibilities for use in cropping systems (Katarzyna et *al.*, 2018) Firstly, camelina can be used as a main crop on marginal land (C. Zhang & Auer, 2020) . This not only allows the use of undeveloped land, but also improves the overall soil quality (Zanetti, Alberghini, Marjanović Jeromela, et *al.*, 2021) . In addition, camelina can also be used as a cover crop in double cropping systems. In this system, camelina is planted as a cover crop between the main crops to provide ground cover, reduce weed growth and prevent erosion during the non-cropped season by improving water infiltration (Gaba et al., 2015; Meyer et al., 2019) . In addition, camelina can be intercropped with other plants such as legumes or grasses (Zanetti, Alberghini, Marjanović Jeromela, et al., 2021) . This association system offers several advantages, such as improved soil fertility thanks to nitrogen fixation by legumes, increased biodiversity and the possibility of diversifying income through the sale of several crops. In addition, camelina can also be used as a relay crop in cropping systems (Berti et al., 2015) . In relay cropping, camelina is planted in the same field after the main crop has been harvested. This makes efficient use of resources and extends the growing season, maximising productivity on the same land. These various uses of camelina offer farmers opportunities to improve soil health, increase biodiversity and diversify income sources, (Sindelar et al., 2017; Struik & Kuyper, 2017) .

Camelina's remarkable agronomic plasticity means that it can be easily integrated into a variety of cropping systems, in both conventional and organic farming (Bilalis et al., 2017; Royo-Esnal & Valencia-Gredilla, 2018) . Mixed cropping, where several crops are grown simultaneously in the same field, has become common practice, particularly in organic farming, where it provides effective control of weed pressure (Gollner et al., 2010; M. Leclère et al., 2019; SAUCKE & ACKERMANN, 2006) . Camelina is also widely used as a cover crop, a growing trend due to its undeniable environmental benefits (Petit et al., 2018) , offering camelina the opportunity to move from a niche crop to a profitable crop (Zanetti, Alberghini, Marjanović Jeromela, et al., 2021) .Several studies have demonstrated the promising results of camelina as an association crop , for example by combining it with strawberries in winter cultivation (Groeneveld & Klein, 2014) or with maize in summer cultivation (Zanetti et al., 2017) . Mixed cropping is of crucial importance for producers in regions facing water scarcity constraints (Gaba et al., 2015) , offering a significant opportunity to make the most of resources by growing two species at the same time. Recent research has shown the success of camelina in association with lentils or peas (M. Leclère et al., 2019; Paulsen, 2011) , with no significant yield loss and no need for herbicide intervention. Promising trials in Austria Gollner et al., (2010) have also been carried out combining camelina with soya and lentils. However, strong competition with grasses, mainly wheat and barley, can lead to a significant reduction in camelina yield (M. Leclère et al., 2019; Paulsen, 2011) .

5.3. SWOT analysis of Camelina (*Camelina sativa*) :

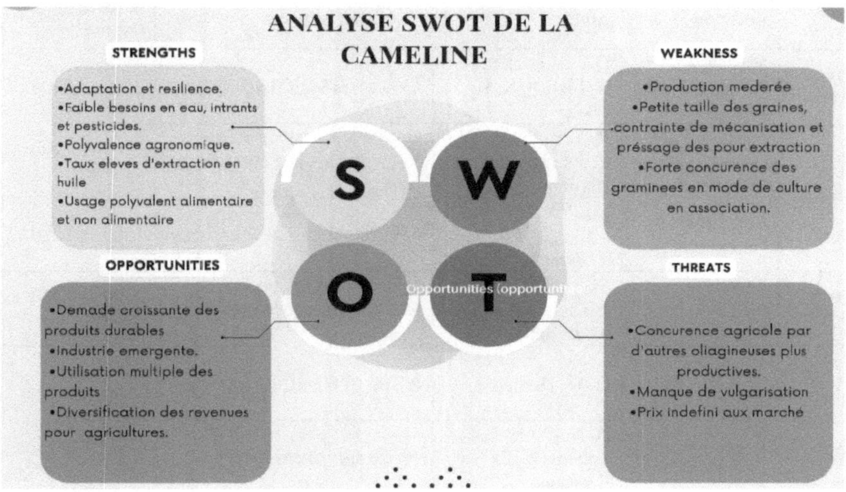

Figure 7. SWOT analysis of Cameline

Conclusion:

The SWOT analysis of camelina highlights its strengths in terms of adaptability, low water consumption and agronomic versatility. However, challenges such as moderate productivity and competition with other crops need to be addressed. The opportunities offered by the growing demand for sustainable products and emerging industries can compensate for these weaknesses. Farmers and industry players must remain alert to threats such as agricultural competition and price fluctuations. By adopting sustainable farming practices and exploring various applications, camelina can play a significant role in promoting environmental and economic sustainability in Algeria.

6. Camelina productivity under limited water supply

The impact of deficit irrigation on Camelina productivity has been studied in several research studies. These studies have explored the effects of different levels of water supply on Camelina growth, development and yield. The following is an overview of the main research carried out in this field:

Table 2. Summary of Studies (2008-2021) on the Impact of Different Irrigation Regimes on Different Yield Components in Different Environments.

Year	Location	Authors	Irrigation schemes	Seed yield (kg/ha)
2008-2010	Maricopa, Arizona, United States	D. J. Hunsaker, A. N. French, K. R. Thorp	45% vs. 60% of field capacity	1500 vs 1200
2011	Klamath Falls, Oregon, United States	R. S. B. Smith, W. C. Johnson, J. L. Hatfield	Once a week vs. once every fortnight vs. once every three weeks	1500 vs 1200 vs 1000
2012	Reno, Nevada, United States	J. E. Thomas, R. D. Miller, D. M. Anderson	45% vs. 60% of field capacity	1500 vs 1200
2013-2017	Several sites in the United States	M. A. E. B. Silva, J. Thomas, D. M. Anderson	Once a week vs. once every fortnight vs. once every three weeks	1500 vs 1200 vs 1000
2014	Maricopa, Arizona, United States	G. W. Hergert	0, 100, 200 and 300 mm	1000, 1200, 1500 and 1800
2015	Reno, Nevada, United States	D. Neupane, R. H. Lohaus, K. F. Bronson	10, 20 and 30 cm	1000, 1200 and 1500
2016	Reno, Nevada, United States	A. D. Pavlista	0, 10, 20 and 30 cm	1000, 1200 and 1500
2018	California, United States	Nicholas George	100 and 200 mm	1000 and 1500
2019	Maricopa, Arizona, United States	K. F. Bronson	0.5-1.1 irrigation fraction	1000-1800
2020	Iran	Safoora Borzoo	100, 75 and 50% of field capacity	1500, 1200 and 1000

2020	Pakistan	Muhammad Aown Sammar Raza	0, 1, 2 and 3 irrigations	1000, 1200, 1500 and 1800
2021	Nevada, United States	Zeeshan Ahmed	0, 1, 2 and 3 irrigations	1000, 1200, 1500 and 1800

The distribution of the 12 studies selected over a period extending from 2008 to 2021 highlights the growing interest and expansion of camelina research over the last decade. Several factors may explain this trend. Firstly, there is a search for alternative crops, particularly in regions facing challenges linked to water availability. Camelina has the advantage of a reduced environmental footprint and significant economic value, helping to reduce dependence on fossil fuels while offering a profitable agricultural alternative, particularly in regions where water is scarce.

Geographical breakdown :

An analysis of the geographical distribution of the 12 studies on the effect of irrigation on Camelina yield shows a clear concentration in the United States. The majority of the studies (10 out of 12) were carried out in the United States. This concentration can probably be explained by several factors, including :

The strong interest shown by American researchers in alternative and potentially sustainable crops. The presence of arid and semi-arid zones in the United States, where Camelina could play an important role in sustainable agricultural production.

Only 2 studies have been carried out outside the United States: Pakistan (1 study) and Iran (1 study).

Geographical diversity is therefore limited for the time being, but with the growing interest in Camelina, we can expect to see more studies in other regions of the world, particularly in arid and semi-arid zones. The absence of European countries in the studies is also consistent with this hypothesis. Europe is a region where the climate is generally more temperate, with more abundant rainfall. Under these conditions, Camelina is less likely to suffer from water stress, and irrigation may therefore be less necessary.

Cultural aspects of camelina Under deficit irrigation :

In addition to the information summarised above, the studies we have provided also contain more specific data on various aspects of growing camelina under deficit irrigation. For example:

The effect of different irrigation rates on yield, oil content and fatty acid composition of camelina seeds (studies N°1, N°4, N°6, N°8, N°9, N°10, N°11, N°12)

The effect of nitrogen fertilisation on the yield and oil content of camelina (study N°6).

Comparison of the performance of camelina and canola under deficit irrigation (studies N°2, N°8)

The use of simulation models to study camel water consumption (study No. 9).

Research on Camelina under deficit irrigation has highlighted its ability to tolerate water stress conditions while maintaining a certain level of productivity. Studies have shown that Camelina has physiological and morphological mechanisms that enable it to adapt to limited water levels, by modifying its transpiration and water use patterns to optimise its yield under stress conditions.

The research also examined the effect of different deficit irrigation regimes on the growth and development of Camelina. The results indicate that reduced levels of water supply during certain growth phases can lead to a reduction in seed yield. However, Camelina shows some resilience to these conditions, with adjustments in its metabolism and physiology to maintain productivity in water-limited environments.

The researchers also examined agronomic practices that could improve the performance of Camelina under deficit irrigation. Strategies such as fertiliser management and the selection of water-stress resistant varieties to mitigate the negative effects of water shortage on the crop.

In summary, the available studies suggest that Camelina has some ability to thrive under deficit irrigation conditions. However, further research is needed to better understand the mechanisms underlying this tolerance to water stress and to develop appropriate management strategies to optimise Camelina productivity in environments where water resources are limited.

Recommendations

There is no doubt that further research into new oilseed crops, particularly with regard to water requirements and water stress management, is crucial to making these crops economically viable in semi-arid environments such as Algeria. A precise understanding of the water requirements of these crops and their ability to withstand water stress is essential to optimise their yield and efficiency in such conditions.

One of the less explored areas in oilseed research is irrigation and its impact on various production parameters. Studies on the efficient use of irrigation, including irrigation techniques and irrigation strategies adopted, can provide valuable information on how to optimise the growth and yield of oilseeds while making efficient use of available water resources.

In addition, in-depth research into the effects of irrigation on the quality of oilseed crops, such as oil content, fatty acid composition and seed quality, is also needed to guarantee the profitability and commercial value of the crops.

By investing in research into irrigation and its impact on oilseeds, Algeria and other semi-arid regions can develop more sustainable and profitable farming practices, reduce their dependence on rainfall and maximise their agricultural production in difficult climatic conditions.

Chapter II. Materials and methods

1. Characteristics of the experimental site :

The experiment was carried out over two cropping seasons, 2021-2022 and 2022-2023, at the experimental station of the Institute of Agronomic and Veterinary Sciences of Batna 1 University, located in central-eastern Algeria (Figure 8). The geographical coordinates of the station are 35° 32' 41" N and 6° 8' 56" E, with an altitude of 1050 metres. The region belongs to the semi-arid bioclimatic zone. The study area is characterised by hot, dry summers and cold, relatively wet winters.

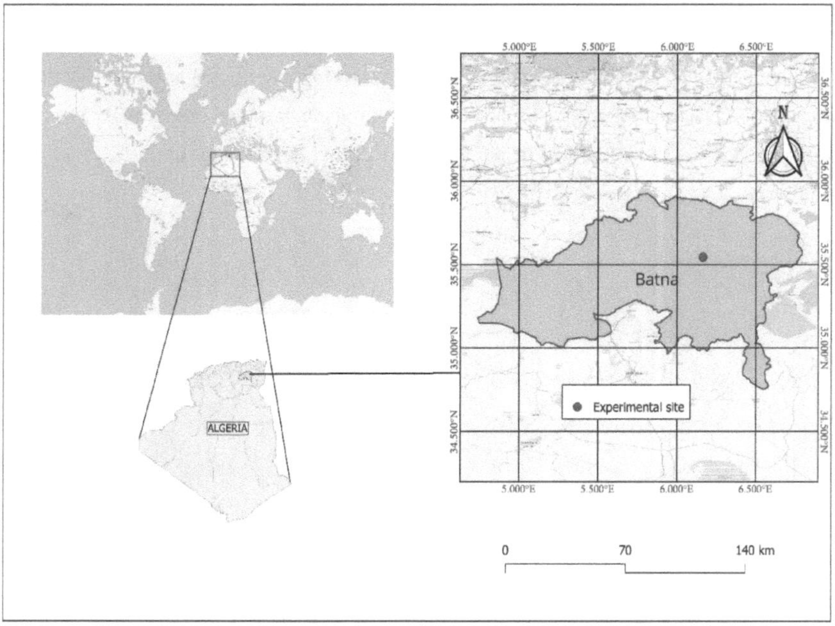

Figure 8. Geographical location of the study area

The soil has a fine silty texture, with a high fine silt content of 58% and a low organic matter content of 0.7%. The soil analyses were carried out in the laboratory of the Centre de Recherche Scientifique et Technique sur les Régions Arides (CRSTRA), in accordance with the soil analysis methods and procedures (Table 3) defined by the International Soil Reference and Information Centre (ISRIC), following the methodology of Van Reeuwijk, (1995).

Table 3. Values of soil characteristics and properties

The parameter	Depth of 0-60cm
Electrical conductivity ds/cm	0.21
PH	8.37
$Caco_3$(Total limestone) (%)	36.75
Organic matter (%)	0.74
$P_{(2)}o_5$(meq/100g)	17.16
Ca^{++} (meq/100g)	6.45
Na^{+} (meq/100g)	3.83
K^{+} (meq/100g)	0.368
N nitrogen (%)	0.42
Carbon (%)	2.50
Clay (%)	20
Fine silt (%)	58
Coarse sand (%)	05
Fine sand (%)	10
Coarse silt (%)	07
Humidity at wilting point PF (%)	34.5
Moisture at cc field capacity (%)	18.7
Bulk density (g/cm^3)	1.30
Texture	Fine loam
Battance index	3.16

The soil has a high capping index, calculated according to the Rémy and Marin-Laflèche (1974) formula, which is equivalent to a significant risk of water and air impermeability. To remedy this situation, we recommended adding organic matter in the form of cattle manure, at a rate of 3 kg per square metre respectively. This correction resulted in a satisfactory reduction in the battance index to a satisfactory value of 1.38, while the organic matter content rose to 2.3%.

2. Setting up the trial and growing the crop :

2.1. Plant material and experimental set-up.

The Alba genotype of *Camelina sativa* was broadcast at a density of 4 kg ha^{-1}, i.e. approximately 400 seeds per square metre. The seed was sown to a depth of 10 mm on 2 January 2022, after a crop of durum wheat in the first year, followed by oilseed rape in the second year. The second year was sown on 25 November 2023. The seeds, supplied by the Institut National de la Recherche Agronomique (INRA) in Setif, had a thousand grain weight (MGW) of 1.01g. The germination capacity (95%) was determined ten days before sowing according to the ISTA protocol (International Seed Testing Association, 2021) at the Centre National de Certification des Semences de Sétif (CNCC).

We chose a totally randomised system with four irrigation regimes repeated four times, namely :

T0: Control treatment (Rain)

T1: Treatment with full irrigation covering all water requirements 100% ETc

T2: Treatment with irrigation covering 75% ETc

T3: Treatment with irrigation covering 35% ETc

The individual plots were square in shape, measuring 6.25 m² in area, 2.5 m long and wide, with a spacing of one metre between them.

Figure 9. Experimental set-up

2. 2. management and cultivation

2.2.1. Irrigation management

For the management of water supplies in our study, we opted for sustained deficit irrigation rather than regular deficit irrigation based on water restrictions during well-defined short periods. This decision was motivated by the advantages of sustained deficit irrigation in terms of saving water and maintaining yields. Unlike the phenological model based on growth stages, sustained deficit irrigation involves water restrictions throughout the growth cycle. A recent study by Ouda et *al.*, (2021) showed that sustained deficit irrigation was more suitable than the phenological model in terms of reducing yield losses and saving irrigation water. This is a relevant approach in the current context of increasing pressure on water resources and trends linked to climate change.

The various irrigation requirements were calculated using three lysimeter boxes equipped with drainage systems, installed near the experimental plots.

Maximum crop evapotranspiration (ETc) and net irrigation quantities were estimated as follows:

$$ET_c = ET_0 \times K_c$$

ET_0: Potential evapotranspiration was calculated experimentally using lysimetric boxes with drainage.

Kc : The cultural coefficient was determined on the basis of phenological stages (D. Hunsaker et al., 2011) .

Reference evapotranspiration (ET_0) was measured over a given time interval using three drainage lysimeters, each with a surface area of 1 square metre (1 m x 1 m) and a depth of 1 metre, spaced 0.5 metres apart. This three-repeat arrangement ensures greater accuracy of measurement. The grass inside the lysimeters was maintained at a height of 15 cm and regularly irrigated. Drainage water was collected and measured in a calibrated container at the bottom of each lysimeter.

Potential evapotranspiration was calculated using the water balance method, as described by Allen et *al*., 1998. It is based on the difference between incoming and outgoing water flows over a given period of time and is expressed as follows:

$$ET_0 = P + I - D \pm \Delta W$$

ET_0: Potential evapotranspiration.

P: Precipitation.

I: Irrigation .

D: drainage water collected at the bottom of the lysimeter.

ΔW: The change in soil water content within the isolated lysimeter remained consistently at field capacity once drainage ceased. ΔW becomes negligible, and the formula for calculating evapotranspiration becomes consistent with Aboukhaled et *al* .., (1982)

$$ET_0 = P + I - D$$

The evapotranspiration values obtained are averages calculated over the periods between drainage episodes (Aboukhaled et al., 1982) .

2.2.2. Management of predators and weeds

Weeding was carried out manually, periodically checking for weeds.

Unlike oilseed rape, birds and flea beetles were not considered to be pests of camelina (Pavlista et al., 2011) , but the experiment paid particular attention to the presence of other camelina pests, notably ants (Messor barbarus). This species has been observed to cause damage to camelina siliques at the ripening stage, cutting off the petioles and causing the siliques to fall prematurely. In order to prevent yield losses, a chemical intervention was carried out using an insecticide based on *(Pirimiphos-methyl)* in accordance with local regulations and good agricultural practice. This control measure limited the damage caused by the ants and preserved the quality and yield of the camelina crops.

2.2.3. Input management

The amount of fertiliser applied before sowing as a basal fertiliser was 1quintal/ha of *mono ammonium phosphate* MAP (12.52), and as a cover fertiliser the amount of urea 46% used was 50Kg/ha applied at the rosette stage.

2-2-4 Harvesting :

Figure 10. Harvesting camelina.

Harvesting was carried out manually when seed moisture reached between 8 and 10% in order to minimise potential losses associated with late harvesting, such as silique splitting due to wind, rain or even hail.

3. The parameters studied

3.1 GDD (growing degree days)

The GDD parameter, or growing degree days, occupies a central place in modern agriculture as an essential tool for assessing the development cycle of crops. By calculating the sum of temperatures above a certain base temperature, usually expressed in degrees Celsius (McMaster & Wilhelm, 1997) , GDDs can be used to monitor the thermal evolution required for crop growth. This monitoring of GDD accumulation offers farmers and researchers the possibility of accurately predicting critical stages in crop development, such as flowering, fruit set and harvest.

Thanks to the information provided by GDDs, players in the agricultural sector can plan their farming activities more effectively and make informed decisions. The DDGs make it easier to plan sowing dates, crop management and pest control, thereby contributing to better farm management.

In addition to its practical usefulness, the GDD parameter plays a crucial role in raising awareness of the challenges posed by climate change on agricultural production. By monitoring the accumulation of GDD over the seasons, it is possible to identify the effects of climate variations on crop development and anticipate the need for resilient cultivars.

In short, the GDD parameter is a valuable and versatile tool in the arsenal of modern agricultural practices. It provides an in-depth understanding of the crop development cycle and is a major asset for strategic decision-making in agriculture. This parameter is calculated using the formula from McMaster & Wilhelm, (1997) .

$$GDD = \sum_{i}^{j} \frac{\text{Tmax} + \text{Tmin}}{2} - \text{Tbase}$$

TMax: The maximum temperature for the day.

Tmin: The minimum temperature for the day.

Tbase: Since 5°C is the appropriate base temperature for growing camelina, it is used to calculate the GDDs (Aiken et al., 2015) .

3.2. Morphological characteristics

Various physiological growth parameters are measured for each plot by randomly sampling six plants from each plot. This includes determining grain yield, biomass and oil content. A sample of one square metre in the central part of each plot was taken on the day of harvest.

3.2.1. Plant height (HP)

The height (cm) is measured at harvest, from the ground to the tip of the plant.

3.2.2. The number of branches per plant (NBP)

The number of plant branches for each treatment was counted in each elementary plot by randomly sampling ten plants.

3.3. Fertility variable

3.3.1. Thousand kernel weight (TKW)

Because camelina seeds are so small, it is essential to use a professional NUMEGRAL grain counter at the Batna Cereals and Pulses Cooperative (CCLS). The seeds are counted using this specialised equipment, repeated four times to ensure accurate weighing of the number of seeds. The total weight of the grains is then measured using a precision balance SARTORIUS MA 30

.

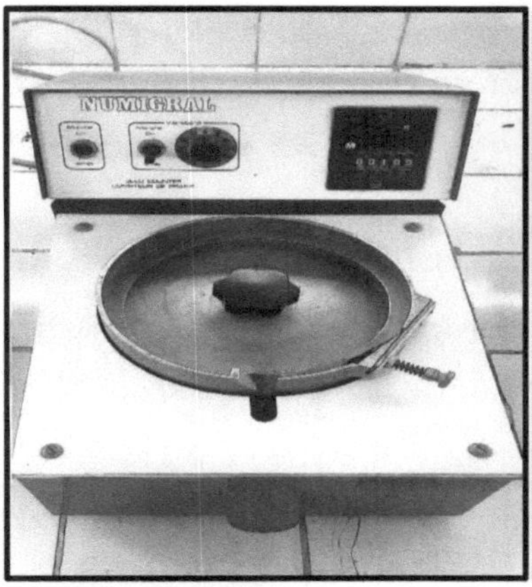

Figure11. NUMIGRAL grain counter **Figure12.** SARTORIUS MA 30 precision balance

3.3.2. Number of seeds per silique (NGS)

The number of seeds per silique is obtained by threshing 10 pods per three plants, taken at random from each elementary plot.

3.4. Crop productivity parameters :

3.4.1. Biomass yield: (qx/ha)

Biomass yield is determined by weighing the harvest from one square metre of the central part of the elementary plot, expressed in grams per square metre (g/m^2), then converted into quintals per hectare (qx/ha).

Figure 13. Weighing a square metre of biomass

3.4.2. Grain yield (qx/ha)

To determine grain yield while separating seeds from straw and pod husks when harvesting a square metre. Expressed in grams per square metre (g/m^2), then converted into quintals per hectare (qx/ha).

3.4.3. Straw yield (qx/ha)

The straw yield (g/m^2) can be determined by subtracting the grain yield from the biomass yield, then converted into quintals per hectare (qx/ha).

3.4.4 Harvest index (%)

The (IR%) is determined as the ratio between the seed yield and the total biomass yield (Donald & Hamblin, 1976) .

3.5. Nutrient quality variables

3.5.1. Protein content (%)

the total nitrogen content is determined using the Kjeldahl method, which involves the following stages: digestion, neutralisation, distillation, dosage and finally calculation of the protein content.

3.5.2. Oil content (%)

As far as the oil content is concerned, it is determined as the corresponding percentage between the mass of oil obtained and the total mass of the seeds, in accordance with the method described by Belayneh et *al.*, (2015) . It should be noted that the pressing is carried out cold, a method which involves extracting the oil from the seeds without the application of heat, thus preserving the quality and properties of the extracted oil as much as possible.

Figure.14 Extraction of camelina oil.

3.6. Energy and oil potential variables

3.6.1. Oil yield (Kg/ha)

The oil yield was calculated as the product of the camelina seed yield multiplied by the percentage oil content of the seeds.

3.6.2. Camelina biodiesel productivity component (L/ha)

The methodology used to estimate biodiesel production was based on the guidelines described by Kemp, (2006) , which involve the use of a mechanical pressing process. This process assumes an extraction rate of around 80%, as opposed to the chemical method of hexane extraction, which makes the by-products too dry to be used as animal feed. Because of the small size of camelina seeds and their susceptibility to post-harvest losses, we assumed a post-harvest loss rate of 10%. The remaining 90% was used to estimate oil yield based on oil content (Neupane et al., 2020; Sintim et al., 2015) . This oil yield was then converted to biodiesel volume, at a rate of 1 kg/ha equivalent to 0.439 L (Kemp, 2006) .

3.7. Forage yield components

3.7.1. Chemical composition of camelina straw and oilcake

The chemical composition of the straw and oilcake was determined by taking several samples during the collection of each elementary plot for the straw, and during the extraction of the oil, the residues were collected at the intermediate stage of extraction to avoid mixing between the extractions of the different treatments. To calculate dry matter (DM), the samples were placed in an oven at 104°C for 4 hours until a constant weight was obtained. The samples were then ground to determine the cellulose (CB), total nitrogenous matter (MAT), mineral matter (MM) and fat content (MG). Conventional chemical analyses were carried out using the official methods of the Association of Official Agricultural Chemists (AOAC) (1990).

Dry matter (DM) (%)

The dry matter is calculated by subtracting the initial weight of the sample from the final weight after oven drying at 104°C for 4 hours. This result is then divided by the initial weight and multiplied by 100 to obtain the percentage of dry matter.

Mineral and organic matter (OM) (%)

The mineral content of a sample is determined gravimetrically after calcination at 550°C for 6 to 7 hours. This process eliminates the organic matter, leaving behind the mineral components in the form of ash. The ash obtained, which must be white or grey with no carbon particles, represents the fraction of mineral matter in the sample.

Total nitrogen matter (%)

Total nitrogen is determined by the Kjeldahl method, which is carried out in three stages: mineralisation, distillation and finally titration.

Crude cellulose (CB) (%)

Crude cellulose was determined as a percentage of the organic residue using the classic Weende method. This method involves two successive hydrolyses: first in an acid medium with sulphuric acid (H_2SO_4, 0.26 N), then in a basic medium with potassium hydroxide (KOH, 0.23 N). The first hydrolysis eliminates acid-soluble carbohydrates and hemicelluloses, while the second hydrolysis in a basic medium eliminates lignins and proteins. The final residue, mainly made up of cellulose fibres, is dried and weighed to determine the crude cellulose content, expressed as a percentage.

3.7.2. Enzymatic digestibility

Enzymatic digestibility was determined using the pepsin-cellulase enzyme method proposed by Aufrere and Doreau (1983) and described by Demarquilly and Andrieu (1978). This method involves the degradation of 0.5g powder samples by two enzymes diluted in appropriate buffers. The first attack is carried out by mixing pepsin (50 ml) for 24 hours incubation in a water bath at 40°C. After filtration and rinsing with distilled water, the residue was attacked with cellulase (50 ml) for 24 hours incubation at 40°C. The residue was then filtered, rinsed and placed in an oven at 104°C for 48 hours. This yields the indigestible dry matter. After weighing, the residue is incinerated in a muffle furnace for 4 hours at 500°C, then weighed. The ash obtained, which corresponds to the mineral matter, is subtracted from the indigestible dry matter to obtain the indigestible organic matter.

The calculation of various parameters, such as the cellulasic digestibility of organic matter in % (DcellMO) and the digestibility of organic matter (DMO), is estimated using the formulas of Demarquilly and Andrien (1987).

$$DMO = 0.656.DcellMO + 24.33$$

0.656 and 24.33 of the constants established by Aufrere and Dominique (1989) for dry fodder.

3.7.3. Calculation of fodder value

The fodder value expressed in fodder units is calculated using the Breirem formula (INRA, 1978), which is a function of digestible organic matter (DOM) and non-digestible organic matter (NODM).

$$MOD = MO.DMO$$

$$MOND = MO - MOD$$

$$UF = 2.36MOD - 1.20\frac{MOND}{1650}$$

FU: forage unit per kg of dry matter.

3.8. Components related to water use efficiency

To assess the efficiency of total water use, we refer to the Water Use Efficiency (WUE) index. This index measures the ratio between agricultural production, expressed as seed yield or total biomass, and the total quantity of water in cubic metres or mm. The calculation of this index follows the methodology detailed by Payero et *al*,(2008).

3.8.1. Total water use efficiency (Kg/m^3)

$EUE_{grain} = (Grain\ yield\ (kg))/(Total\ water\ applied\ (m3))$

$EUE_{Organic} = \frac{\text{Biomass yield (kg) (Grain yield (kg)) / (Total water applied (m}^3\text{))}}{\text{Total water applied (m3)}}$

The total water applied includes both natural rainfall and irrigation water.

3.8.2. Irrigation water use efficiency (Kg/m^3)

$EUEI_{grain} = \frac{\text{Irrigated grain yield - Rainfall grain yield (kg)}}{\text{Irrigation water applied (m3)}}$

$EUEI_{Bio} = \frac{\text{Irrigated biomass yield} - \text{Rainfall biomass yield (kg)}}{\text{Irrigation water applied (m3)}}$

Another important index of water use efficiency measures water productivity per forage unit (EUEF).

3.9 Water use efficiency per forage unit (FU/mm)

3.9.1. Total water use efficiency per forage unit (FU/mm)

$$\text{EUEF} = \frac{Irrigation\ water\ applied\ (m3)\ Feed\ value\ produced\ (UF)}{Total\ water\ applied\ (mm)}$$

The total water applied includes both natural rainfall and irrigation water.

3.9.2 Irrigation water use efficiency per forage unit (UF/mm)

$$\text{EUEIF} = \frac{forage\ value\ in\ irrigated\ mode - forage\ value\ in\ rain\text{-}fed\ mode\ (UF)}{Irrigation\ water\ applied\ (mm)}$$

Chapter III. Results and discussion

1. Impact of Climatic Factors on the Growth of *Camelina sativa*: Analysis of the 2021-2022 and 2022-2023 Crops

1.1. Introduction

Climatic variations play a key role in agricultural production, affecting both crop growth and the economic viability of farms. This research focuses on the oilseed crop *Camelina sativa*, examining the impact of climatic factors over two separate crop years (2021-2022 and 2022-2023).

In this analysis, we will highlight fluctuations in minimum and maximum temperatures, as well as rainfall, in order to determine how these factors have influenced the growth cycle of camelina. Using indicators such as growing degree days (GDD), we will assess the physiological stages of the crop and the impact of sowing date on plant development. Finally, this section will emphasise the importance of camelina adaptation to the challenges posed by climate change, highlighting the implications for the sustainability of cropping systems in semi-arid regions.

1.2 Influence of climatic factors on *Camelina sativa* cropping seasons

1.2.1. First campaign (2021-2022)

Analysis of the first crop year, which ran from 2 January to the end of May, reveals significant variations in the weather conditions observed throughout the trial period. During this period, the lowest average minimum temperatures were recorded in January (decade 3), reaching a threshold of -4.14°C, likely to have a negative influence on the germination and development of cold-sensitive crops, as highlighted in previous work (Mortlock & Vanderlip, 1989) . At the same time, average maximum temperatures peaked in May, exceeding 28°C, which can potentially induce water stress in crops, particularly damaging in the event of insufficient rainfall. May was identified as the hottest period of the experiment, with mean maximum temperatures exceeding 30°C, which could have a negative impact on silique filling and oil content in Camelina sativa (Righini et al., 2019) . As for rainfall, it was generally low during the winter months, but gradually increased from April onwards, peaking at 31.60 mm in March (Figure 15).

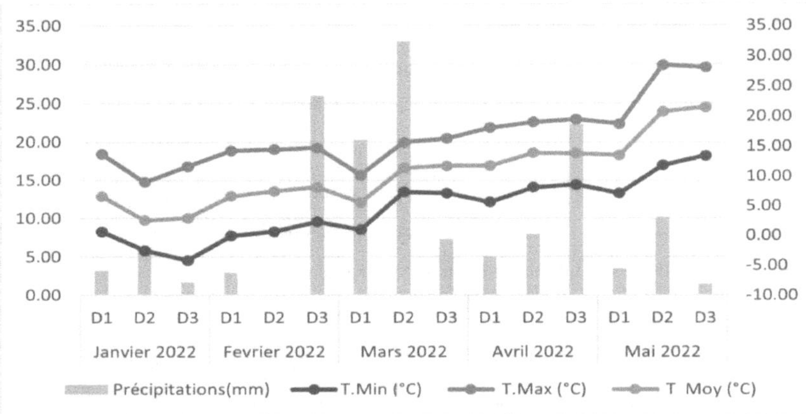

Figure 15: Precipitation and temperatures during the 2021 - 2022 crop year.

1.2.2 Second campaign (2022-2023)

The second crop year, running from 26 November to 12 May, was marked by significant variations in the weather conditions observed over the period. The lowest average minimum temperatures were recorded at the same time as last year in January (dekad 3), with a value of -4.11°C, almost similar to that of the first season. These relatively low temperatures could have an impact on crop growth, particularly for cold-sensitive crops. In contrast, the highest average maximum temperatures were recorded in April (dekad 3), exceeding 26°C, 7.37°C higher than the same period the previous year (Figure 16). This period of high temperatures was associated with a lack of rainfall and unusually strong winds for this trial period, a combination of factors that could affect the growth and filling of camelina siliques, and consequently the final yield.

Precipitation was relatively low for most of the observation period. However, a notable increase was observed in December (dekad 1), with rainfall reaching 16 mm. This variation in rainfall could have important implications for irrigation management and crop development.

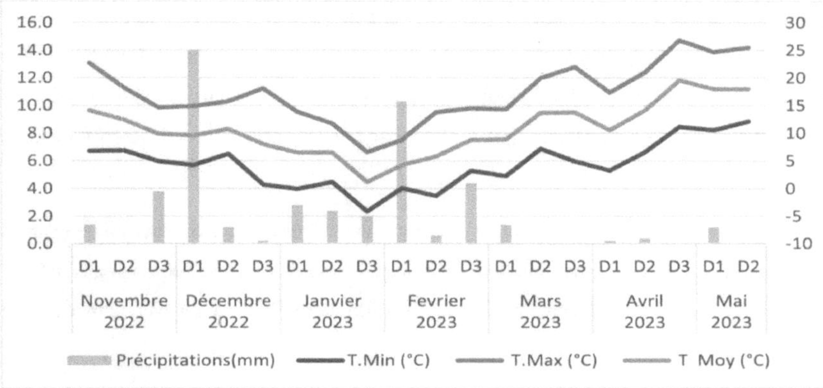

Figure 16. Precipitation and temperatures during the 2022 - 2023 crop year.

1.2.3. Climate summary

Crop management over the two crop years was influenced by two distinct seasons. The first season (2021-2022) was characterised by relatively rainy and favourable conditions, particularly during the critical phases of the crop, when the climate was mild and wet, with 74% of rainfall received during this period compared with the total for the season. In contrast, the second season (2022-2023) was marked by considerably drier and hotter conditions, with NASA even describing it as the hottest and driest year in the history of (Roxana, 2024) . During this period, a total of 44.90 mm of rainfall was recorded between 26 November and 12 May 2023, of which 41.7 mm, or 93% of the total, fell between the sowing date and 1 March. The remainder was spread over the months of March, April and May, which were marked by extreme temperatures. These conditions led to an exceptionally dry year, described as a "blank year" for rain-fed crops in the Batna region. Climate change is disrupting the configuration of food systems by reducing agricultural production. It leads to droughts or floods, and alters the length of the season (Gregory et al., 2005) . Consequently, climate change is a factor leading to a decrease in crop productivity (D. Leclère et al., 2014) .

1.3. Growing Degree Days (GDD)

The camel growth cycle lasted 148 days for the first season, from 2 January 2022 to 30 May 2022, and 168 days for the second season, from 25 November 2022 to 12 May 2023. The length of the development cycle can be expressed in growing degree days (GDD) for the cycle as a whole and for each physiological stage. The main physiological stages have been identified using the three-digit BBCH code, a scale that enables plants to be followed precisely and uniformly throughout their growth cycle, from germination to fruit ripening. Developed by the German Agricultural Research Association (Biologische Bundesanstalt, Bundessortenamt und CHemische Industrie), this three-digit scale is used for a more detailed description of the different stages of crop growth.

These data illustrate the development of the growth stages of camelina over time (Table 4) and make it possible to compare the two crop years in terms of days after sowing (DAS) and growing degree days (GDD) calculated using the formula of McMaster & Wilhelm, (.1997)

Table 4 . Growth phases of Camelina: BBCH, JAS, and GDD codes for the 2021-2022 and 2022-2023 crop years

Growth stage	BBCH code	Cumulative DAS (days after sowing)		GDD	
		2021-2022	2022-2023	2021-2022	2022-2023
Sowing	000	00	00	00	00
Emergence	009	42	18	56.50	88.10
Rosette	202	69	37	59.65	97.60
Start of elongation	301	78	105	64.30	94.70
End of elongation	309	95	112	123.60	105.85
Appearance of flower buds	505	101	116	60.50	30.75
10% open flowers	601	106	119	36.20	24.10
Fruit set	609	118	133	112.35	104.15
All siliques reach final size	709	131	143	115.35	84.75
Siliques ripe for harvesting	809	142	161	192.85	233.10
Harvesting	909	148	168	83.75	98.05
Cumulative		**148 days**	**168 days**	**905 °C**	**961.15 °C**

According to the results, the time taken to reach the emergence stage (009) was 42 days JAS for the first campaign, while it only took 18 days for the second campaign. This difference can be attributed to the different sowing dates between the two campaigns. For the first campaign, sowing took place on 2 January, coinciding with very low temperatures in January and February (-6.10°C), which delayed germination of the camelina seeds. In contrast, for the second campaign, sowing was carried out on 25 November 2022, in more favourable temperature conditions, which enabled faster emergence. This underlines the importance of the sowing date in crop development. Camelina showed remarkable resistance and tolerance to frost and cold.

After the rosette stage (202), weather conditions in the first season were favourable, with cool temperatures (12°C) and sufficient rainfall, especially between the early elongation and fruit set stages, when the crop benefited from 83.1 mm of rainfall, i.e. 55% of the total rainfall. In the second season, however, the crop suffered from severe drought, combined with an exceptional rise in temperatures to 31°C and a very low cumulative rainfall of 1.6 mm, which seriously affected flowering, the number of siliques and, consequently, the final yield and oil content for the rainfed treatment. The anticipated effects of climate change could lead to a drop in agricultural production of more than 57% (Asseng et al., 2015) .

The GDD values calculated for the two seasons are lower than those reported in other studies carried out by Righini et *al*, (2019) in Italy, with a GDD of 982°C, and by Zanetti et *al*, (2017) in Poland with a GDD of 930°C. The fact that camelina has a short growth cycle (just over 900°C) makes it a potentially interesting replacement crop for failed winter crops. Its advantages include rapid crop rotation, enabling farmers to adapt their farming practices more quickly and diversify their sources of income. In addition, the short growth cycle allows greater flexibility in crop rotation, which can contribute to the sustainability and diversification of farming systems.

2. Response of the oilseed crop *Camelina sativa* to the Limited Water Input Monitoring System, Conducted in the Semi-Arid Environment of Aurès

In this section, we will examine the effects of different irrigation treatments on camelina cultivation during the two successive crop years 2021-2022 and 2022-2023. The main objective of this study is to understand how deficit irrigation practice influences water use efficiency, plant growth, crop quality, power consumption, oil content and biofuel potential estimation. In addition, we will analyse the effect of annual variability in climatic conditions on these parameters.

We are particularly interested in the impact of water stress levels on camelina seed productivity and quality, by comparing different growth and yield parameters under different irrigation regimes. This analysis is essential for developing irrigation management strategies that maximise agricultural productivity while conserving water resources and optimising energy efficiency, which is crucial for the sustainability of farming systems in semi-arid environments.

In addition, we explored the relationships between the different variables studied and how they interact with each other using powerful statistical tools such as the correlation matrix, principal component analysis (PCA) and the hierarchical tree. These methods enable us to unravel the complex links between variables and identify the key factors influencing crop growth and yield.

By exploring these various aspects, we aim to provide valuable information that will enable farmers and decision-makers to better understand and manage irrigation for sustainable, high-quality agricultural production. This study promises to reveal fascinating insights into optimising camelina cultivation and water management.

2.1. The effect of deficit irrigation treatments on the crop parameters studied .

The results obtained from our experimentation resulting from treatments of increasing water restrictions over two cropping seasons (2021-2022) and (2022-2023) were processed by analysis of variance ANOVA with two factors (irrigation regimes) and experimental year (Year).

2.1. 1. Morphological characteristics: Plant height (HP), number of branches per plant (NBP) and number of plants per square metre (NPM).

The two-factor analysis of variance revealed significant results concerning the effect of deficit irrigation and year on several morphological measurements of camelina. With regard to plant height (HP), the different treatments showed significant effects ($p < 0.01$) (Figure 17), while year did not have a significant effect. Similarly, the number of branches per plant (NBP) was significantly influenced by the treatments ($p < 0.01$) (Figure 18). However, the effect of the year was not significant. The number of plants per square metre (NPM) was significantly influenced by the treatments ($p < 0.01$), while the year showed no significant effect (Figure 19). These results highlight the significant impact of treatment on camelina morphological variables, while the effect of year was unremarkable (Table 5).

Table 5. Mean squares of variance of the analysis of variance of morphological characteristics of camelina during the two seasons (2021-2022) and (2022-2023) .

Source	Treatment (T)	Year (A)	T x A	Groups
Ddl	3	1	3	
HP (cm)	1219.826**	21,288^{NS}	21,519^{NS}	A(T1),B(T2),C(T3 and T0)
NBP	26.681**	0.011^{NS}	0.371^{NS}	A(T1 and T2),B(T3 and T3)
NPM	145297.083**	16020.50^{NS}	1145.583^{NS}	A(T1 and T2), B(T3),C(T0)

HP: plant height;NBP: number of branches per plant;NPM: number of plants per square metre;Ddl: degree of freedom;NS: not significant;** significance ($p < 0.01$);* significance ($p < 0.05$),

In the context of the agro-climatic circumstances in which the experiment was conducted, it is important to note that the effect of the treatment was predominant on the growth parameters during both campaigns. However, the effect of the experimental year was very limited, even though the second campaign was drier, because the climatic conditions that year were relatively favourable for the installation and start-up of the crop during the first half of the development cycle. This was in contrast to the second half of the development cycle, which is explained by the non-significance of the (year) effect for HP, NPB and NPM. This suggests that the variations observed in the growth parameters are mainly attributable to differences in the treatments applied rather than to seasonal variations between the two years. In this respect, the plants showed an adaptive response by reducing the height and number of branches to compensate for the loss of biomass due to water stress. The results of Ahmed et *al* (2019, 2020) and Mesbah (2009)) support our conclusions.

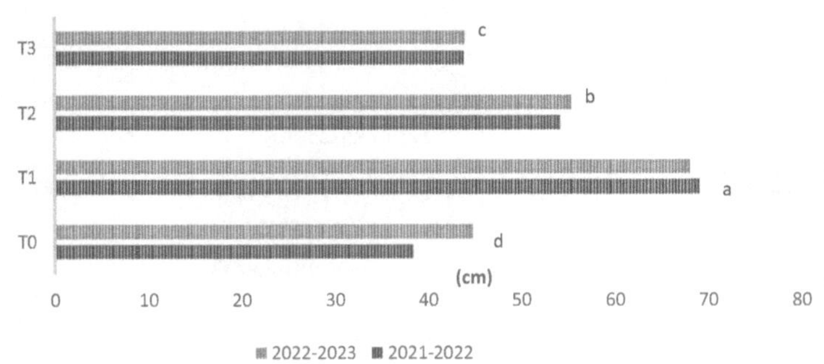

Figure 17. Effect of irrigation regimes on camelina height (HP) during the two seasons (2021-2022) and (2022-2023)

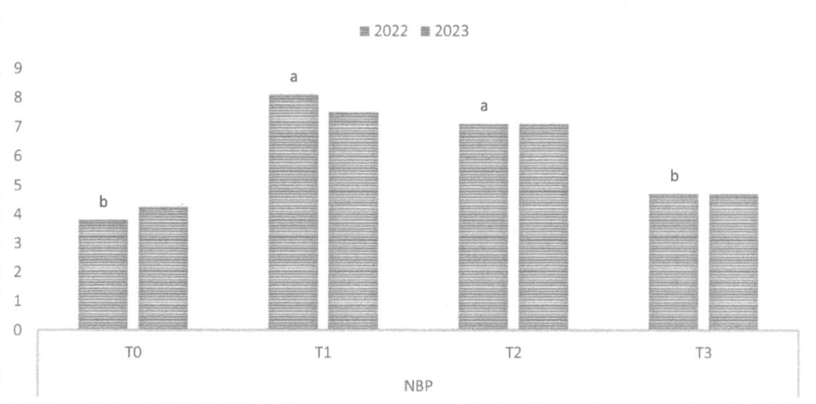

Figure 18. Effect of irrigation regimes on NBP during the two seasons (2021-2022) and (2022-2023)

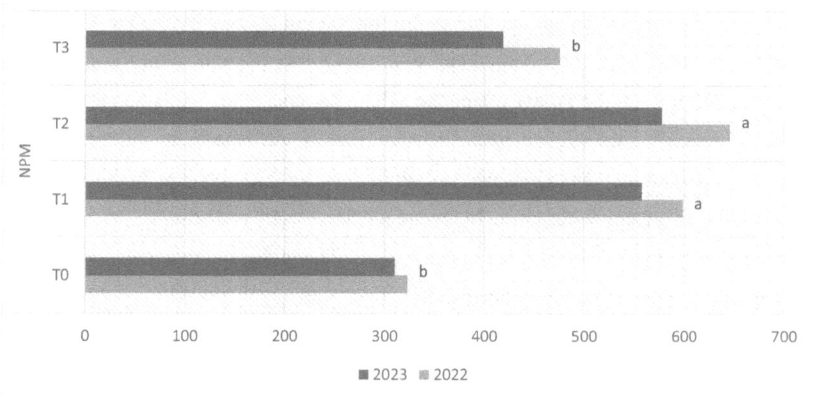

Figure 19. Effect of irrigation regimes on NPM during the two seasons (2021-2022) and (2022-2023)

2.1.2. Fertility variables: number of seeds per silique (NGS) thousand seed weight (PMG)

2.1.2.1. The effect on the number of seeds per silique (NGS)

Analysis of variance (ANOVA) revealed that, unlike morphological characteristics, the number of seeds per silique (NSK) was significantly affected by irrigation treatment, year and their interaction ($p < 0.01$) (Table 6). More specifically, the results show that NGS decreased markedly in the second crop year (2022-2023) compared with the first year (Figure 20).

This drop in NGS in the second season was mainly due to unfavourable weather conditions, in particular the high temperatures and severe drought experienced during the second half of the camelina's development cycle. These conditions particularly affected the flowering and seed filling phases, reducing the plant's ability to produce an optimum number of seeds per silique.

The significant interaction between year and irrigation treatment highlights the importance of climatic conditions in plant fertility.

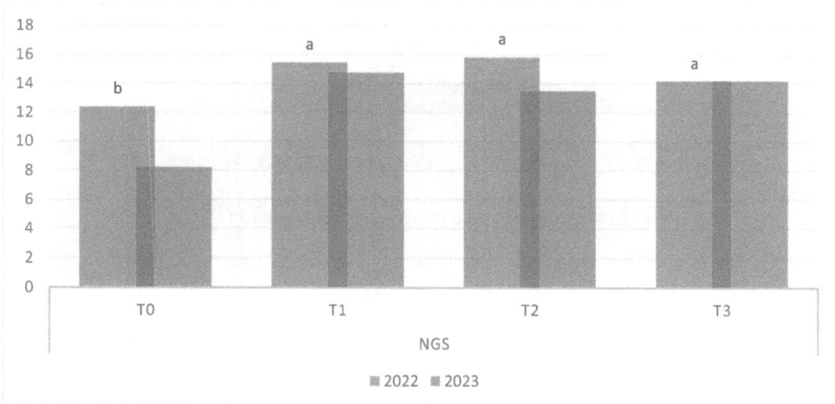

Figure 20: Effect of irrigation regimes on NGS during the two seasons (2021-2022) and (2022-2023)

Table 6. Mean squares of variance analysis of camelina fertility parameters over the two cropping seasons (2021-2022) and (2022-2023).

Source	Treatment	Year	T x A	Groups
Ddl	3	1	3	
NGS	38.637**	25.920**	6.850**	A(T1,T2 and T3), B(T0)
PMG (g)	0.024**	0.047**	0.005^{NS}	A(T1),AB(T2),BC(T3),C(T0)

NGS :Number of seeds per silique ;PMG :Thousand grain weight ; Ddl :degree of freedom ;NS :not significant ;** significance($p < 0.01$) ;*significance($p < 0.05$),

2.1.2.2. The effect on thousand-seed weight (TSP)

Thousand kernel weight (TKW) is a fundamental parameter in the assessment of camelina yield and quality, providing an accurate indication of seed density and size, which are determining factors for overall crop productivity (X. Cheng et al., 2017) .

The impact of deficit irrigation on PMG is less obvious than for other parameters studied (Figure 21). Although a slight downward trend was observed with the restriction of water supplies, the variations remained significant in terms of the effect of treatments and years, although there was no significant interaction between them (Table 6). In particular, the highest value was recorded with treatment T1, followed by treatments T2, T3 and T0.

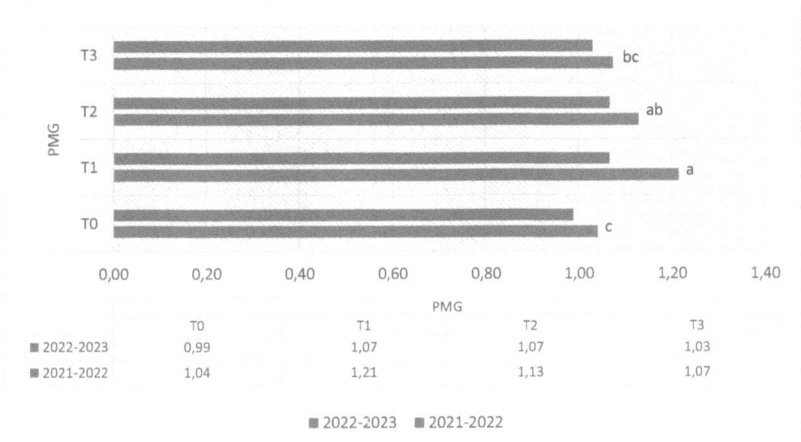

Figure 21: Effect of irrigation regimes on GMP during the two seasons (2021-2022) and (2022-2023).

This observation suggests that camelina is able to maintain a certain grain quality despite water stress induced by different irrigation treatments. Previous studies Ahmed et *al.* (2019, 2020)) have reported that water restriction can disrupt and reduce seed weight. Camelina's small seed size is often considered to be the main constraint limiting its yield compared with other conventional oilseed crops. Improving camelina's GMP is therefore a major challenge for genetic improvement programmes, in order to position this crop not only as a niche crop, but also as one with great potential.

2.1.3. Productivity variables: Grain yield (RDTG), biomass yield (RDT bio) and harvest index (IR)

The response of the productivity components (grain yield, biomass, straw and harvest index) from treatments T0, T1, T2 and T3 are summarised in Table 7.

Table 7. Mean squares of variance analysis of camel productivity parameters over the two cropping seasons (2021-2022) and (2022-2023).

Source	Treatment	Year	T x A	Groups
Ddl	3	1	3	
Organic RDT (qx.ha$^{(-1)}$)	813811.46**	1653.125^{NS}	2461.46^{NS}	A(T1),B(T2),C(T3),D(T0)
RDTG (qx.ha^{-1})	26300.42**	25088.000**	617.250*	A(T1),B(T2),C(T3),D(T0)
RDT Pll (qx.ha^{-1})	5485.002**	138,611^{NS}	34,287^{NS}	A(T1),B(T2),C(T3),D(T0)
IR (%)	.001^{NS}	0.063**	0.000^{NS}	A(T0,T1,T2 and T3)

RDT Bio :Biomass yield ;RDTG :Grain yield ;RDT Pll :Straw yield ; ;IR :Harvest index ; Ddl :Degree of freedom ;NS :not significant ;** significance($p < 0.01$) ;*significance($p < 0.05$),

2.1.3.1. The effect on grain yield (RDTG), biomass (RDT bio) and straw (RDT Pll) .

The two-factor analysis of variance (ANOVA) revealed significant results concerning the effect of irrigation treatment and year on grain (RDTG), biomass (RDTBio) and straw (RDT Pll) yields of camelina. Firstly, the effect of irrigation treatment is clearly demonstrated by the significant differences observed between the different treatments (Figure 22 and 23). Treatment T1, with full irrigation covering 100% of water requirements, gave the highest grain and biomass yields. There was a significant decrease in grain and biomass yields with increasing water restriction in both crop years, with treatments T2 and T3 showing intermediate yields between T1 and T0, the control treatment, a trend confirmed by the significant groups formed for these treatments (Figure 22 and 23). Similar results have also been reported by Ahmed et *al*, (2020); Amiri-Darban *et al*, (2020); Pavlista et *al*, (2016); Raza et *al*., (2015) which validate our results.

Moreover, the effect of year was also significant, as shown by the interaction between year and treatment (T x A) for grain yield. The variations observed from one year to the next can be attributed to changing climatic conditions, which directly influence crop yields. Indeed, the years 2022 and 2023 have different climatic conditions, which is reflected in the yields obtained. In particular, yields in 2023 are generally lower than in 2022, which could be attributed to less favourable weather conditions during the camelina's growing season.

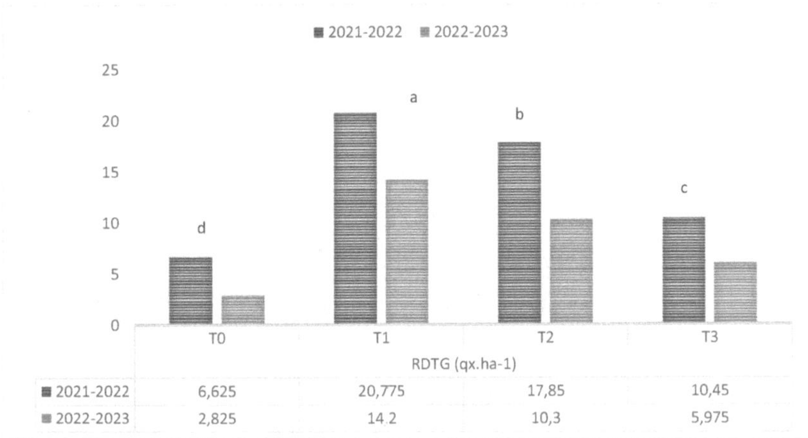

Figure 22. Effect of irrigation regimes on RDTG during the two seasons (2021-2022) and (2022-2023)

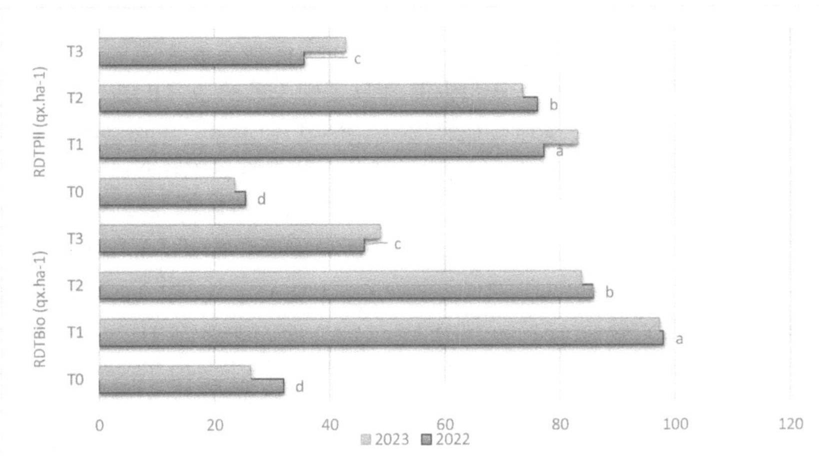

Figure 23. Effect of irrigation regimes on RDT Pll and RDT Bio during the two seasons (2021-2022) and (2022-2023)

2.1.3.2. The effect on the harvest index (HI)

The harvest index, defined as the ratio of grain yield to total biomass weight (Donald & Hamblin, 1976), is a crucial indicator of agricultural productivity, and is of particular importance in assessing the effectiveness of agricultural practices. It is logical to assume that any reduction in biomass yield due to deficient irrigation will also have an impact on grain yield, as these two parameters are often correlated. Furthermore, the small size of camelina seeds can pose additional challenges for measuring the harvest index.

Although there were differences between the treatments, they were not statistically significant and belonged to the same homogeneous group (Figure 24). However, it is interesting to note that treatment T3 showed the best harvest index (17.83%), followed closely by treatment T1 with a slight difference (16.86%), while rainfed treatment T0 recorded the lowest result (15.75%). This observation suggests that even with more restricted irrigation, camelina can maintain a satisfactory RI ratio.

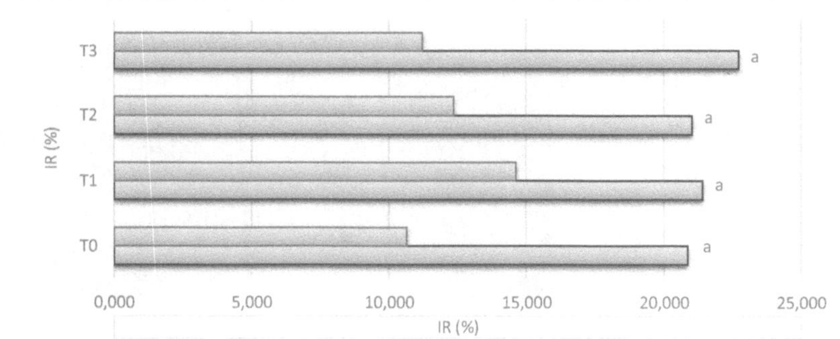

	T0	T1	T2	T3
2022-2023	10,647	14,625	12,360	11,209
2021-2022	20,868	21,409	21,028	22,731

2022-2023 2021-2022

Figure 24. Effect of irrigation regimes on RI during the two seasons (2021-2022) and (2022-2023)

On the other hand, the year effect was significant ($p > 0.01$) with no interaction effect (T x A), with a significant drop in grain yields observed in the second half during the flowering and filling phase in the 2022-2023 season, compared with the first half, which was relatively favourable and provided a relatively acceptable biomass. Our results are consistent with those of Neupane et *al.* (2020)), but contradict those of Ahmed et *al.* (2019)), thus highlighting the complexity of the interactions between agricultural practices, environmental conditions and crop yields.

2.1.4. Variables of Quality of Nutrient Compounds: Protein content (Tprot) and Oil content (Thuile)

2.1.4.1. Effect on seed protein content (Tprot)

Camelina is an oilseed crop for which seed production is the main source of economic interest. While protein content is another important parameter to consider, these parameters can give us an idea of camelina's ability to adapt to limited irrigation conditions. According to the Kjeldahl assay, the protein levels of camelina seed obtained ranged from 27.98% (T1, 2022) to 37.51% (T0, 2023) for both seasons (Figure 25), values often cited in the scientific literature Krzyżaniak et *al.*, (2019); Zajac et *al.*, (2020); Zubr, (2010) ,The statistical factors (Treatment) and (Year) significantly affected protein content with no significant interaction effect between them (Table 8).

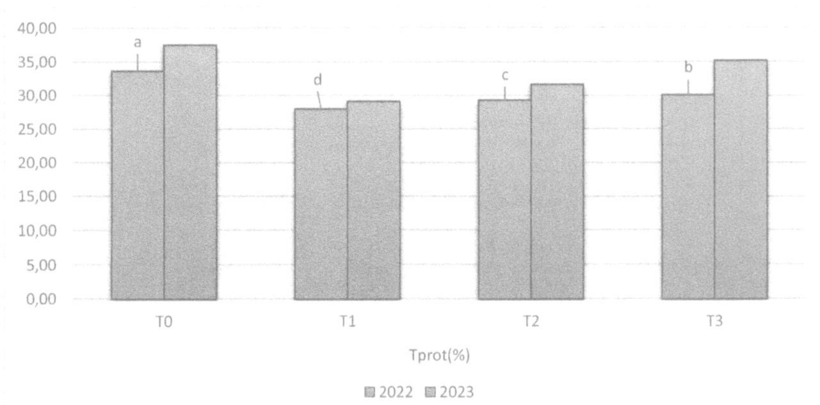

Figure 25: Effect of irrigation regimes on protein content during the two seasons (2021-2022) and (2022-2023).

Table 8. Mean squares of ANOVA variances of camelina Nutrient Compound quality variables over the two cropping seasons (2021-2022) and (2022-2023).

Source	Treatment	Year	T x A	Groups
Ddl	3	1	3	
Tprot (%)	72.872**	77.221**	5,908^{NS}	A(T0),B(T3),C(T2), D(T1)
Thuil (%)	35.781**	52.531**	4,781^{NS}	A(T1),AB(T2), BC(T3),C(T0)

*TH : oil content ;Tprot : protein content :degree of freedom ;NS :not significant ;** significance(p < 0.01) ;*significance(p < 0.05),*

The values show an inverse trend in relation to the other productivity parameters. The protein value increases more and more with water restriction, reflecting the crop's adaptive behaviour to water shortage conditions. They can reduce their metabolic activity and physiological development, resulting in fewer siliques per plant, less branching, reduced thousand-kernel weight and reduced oil content. As a result, a greater fraction of available resources is allocated to protein production, which can lead to an increase in grain protein content in crops grown under water stress conditions (Richard et al., 2000) . The results of this study corroborate the findings of Ahmed et *al*, (2020); Macedo et *al*, (2016); Tari, (2016) , reinforcing our conclusions. In the future, deficit irrigation could lead to significant benefits in yield value through its improvement in seed protein quality (Intrigliolo, 2020) .

2.1.4.2. The effect on oil content (Thuile%)

Oil content is a fundamental parameter for assessing the potential of an oilseed crop. After cold pressing of camelina seeds, the results obtained vary between 23.25% (T0, 2023) and 31.75% (T1, 2022). In particular, treatments T1 and T2 stand out with maximum values for both crop years (Figure 26), averaging 30.50% and 28.87% respectively. In contrast, the T0 rainfed treatment showed a significantly lower oil content, with an average of 27.12% for all crop years. This trend highlights a positive correlation between oil content and water supply, a finding supported by studies by Ahmed et *al*. (2020) and Hergert et *al*. (2016)).

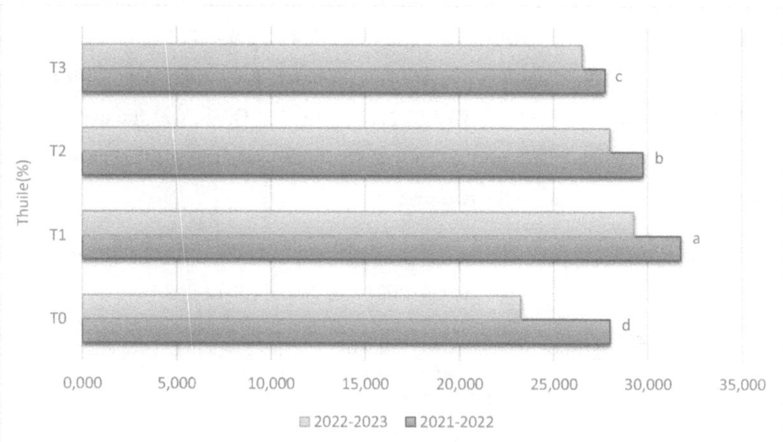

Figure 26. The effect of irrigation regimes on oil content during the two seasons (2021-2022) and (2022-2023)

In addition, the significant effect of the statistical factor (Year) (Table 8) results in a decrease in oil content during the 2022-2023 season, which is marked by drought and high temperatures. This situation has a proportional impact on oil content for all treatments, resulting in a minimum loss of 8% for T1 and a maximum of 17% for T0 (rainfed treatment). These results highlight the crucial importance of water supply on oil yield, as well as the significant impact of unfavourable weather conditions during flowering and seed filling. It is also relevant to note that an accepted oil content can be obtained (T2) even below the maximum values observed for treatment T1 (full irrigation), with a reduction of only 5%, while allowing a considerable saving of 25% in water consumption.

2.1.5. Components related to water use efficiency

Water use efficiency plays a crucial role in sustainable agriculture and the stability of agri-environmental systems. In a context where water resources are increasingly limited and pressure on ecosystems is growing, better water use efficiency enables farmers to reduce their dependence on intensive irrigation. This makes them less vulnerable to water shortages and climate fluctuations, while helping to reduce environmental impacts such as depletion of water resources, soil degradation and pollution.

The challenges of water use efficiency are reflected in the experimental results presented below. The average grain yields for the rainfed treatments for the two seasons were only 4.73 qx. ha-1, with average quantities of irrigation applied for the two seasons: 52.12 mm for T3, 111.69 mm for T2 and 148.92 mm for T1 (full irrigation). The average yield for the two seasons increased to 8.2 qx.ha-$^{(1)}$ (T3), 14.08 qx.ha-1 (T2) and 17.49 qx.ha-1 (T1 full irrigation) (Table 9). In percentage terms, the reduction in grain yield compared with treatment T1 was 72.75% for the rainfed treatment, 53.06% for T3 and 20.77% for T2. Given that the average rainfall during the growing seasons (2022 and 2023) was 98.05 mm, this represented only 39.70% of the average water requirement for the two seasons.

Table 9. Amount of water applied (mm) and water savings under different treatments over the two cropping seasons (2021-2022) and (2022-2023).

Season	Treatment	Irrigation (mm/season)	Effective precipitation (mm/season)	Total water (mm/season)	Relative grain yield (%)	Relative biomass yield (%)	Saving water (%)
(2021-2022)	T_1	136.65	151.20	287.85	100	100	-
	T_2	102.488	151.20	253.688	85.92	87.50	25.00

	T_3	47.828	151.20	199.028	50.30	46.94	65.00
	T_0	-	151.20	151.20	31.89	32.65	100
(2022-2023)	T_1	161.20	44.90	206.10	100	100	-
	T_2	120.90	44.90	165.80	72.53	86.11	25.00
	T_3	56.42	44.90	101.32	42.07	50.12	65.00
	T_0	-	44.90	44.90	19.89	26.99	100

However, the average grain yield EUE values varied between 0.43 $kg.m^{-3}$ (rainfed treatment in the 2022 campaign) and 0.73 kg.m-3 (T2 treatment in the 2023 campaign). Treatment T2 maintained the maximum value for EUEI (which takes into account the difference in yield between irrigated and rainfed grain yield for the amount of irrigation water applied) at 1.095 $kg.m^{-3}$, while the minimum value was recorded for T3, representing a very severe water restriction of 35%, severely impacting grain yield (Fig. 28). This only resulted in an increase of 57.65% and 111.5% compared with T0 for the 2022 and 2023 seasons respectively.

In terms of water use efficiency values for biomass yield, the values for 2023 stand out clearly and are higher than those for 2022 for all treatments (Figure 27). Treatment T0 tops the list with 5.85 kg.m-3 in 2023, twice as much as the previous year (2.08 $kg.m^{-3)}$. However, it is notable that the EUEI values for biomass remain relatively constant from one year to the next for each treatment.

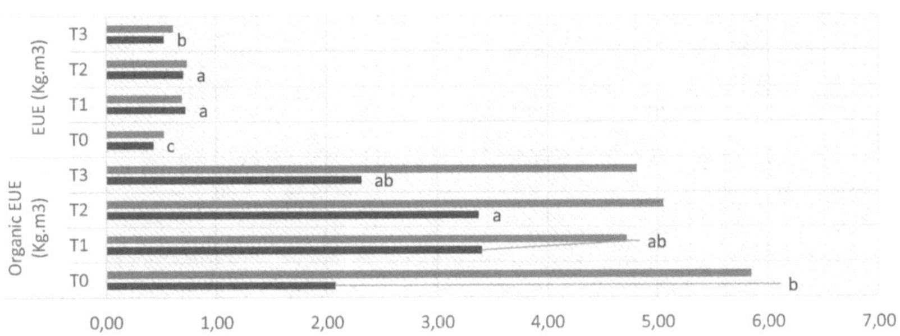

	Organic EUE (Kg.m3)				EUE (Kg.m3)			
	T0	T1	T2	T3	T0	T1	T2	T3
2022-2023	5,85	4,72	5,05	4,81	0,52	0,69	0,73	0,61
2021-2022	2,08	3,40	3,38	2,31	0,43	0,72	0,70	0,53

■ 2022-2023 ■ 2021-2022

Figure 27. Effect of irrigation regimes on grain and organic WUE during the two seasons (2021-2022) and (2022-2023)

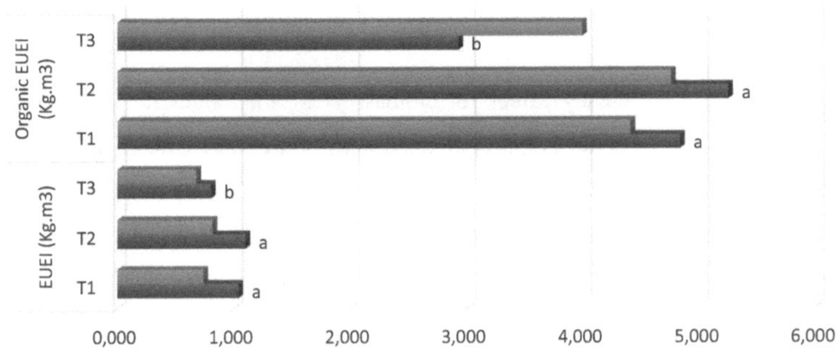

	EUEI (Kg.m3)			Organic EUEI (Kg.m3)		
	T1	T2	T3	T1	T2	T3
2022-2023	0,735	0,813	0,674	4,40	4,76	3,99
2021-2022	1,035	1,095	0,800	4,830	5,245	2,927

■ 2022-2023 ■ 2021-2022

Figure 28. The effect of irrigation regimes on the grain EUEI and organic EUEI during the two seasons (2021-2022) and (2022-2023)

The analysis of variance of the effects of irrigation treatments on crop water use efficiency during the two seasons (2021-2022) and (2022-2023) revealed significant effects ($P<0.01$) of treatments on EUE and EUEI, with a less significant effect ($P<0.05$) for EUEBio and EUEIBio (Table 10). In addition, the effect of agricultural season (Year) was significant ($P<0.01$) for

EUEI and EUEBio, but did not affect the other two variables, EUE and EUEIBio. Finally, the interaction of the two factors (TxA) was only significant for EUEBio. ($P<0.01$)

Average grain yields for the rainfed treatments for both seasons were only 4.73 qx. ha^{-1}, with average irrigation quantities applied for both seasons: 52.12 mm for T3, 111.69 mm for T2 and 148.92 mm for T1 (full irrigation). The average yield for the two seasons increased to 8.21 qx.$ha^{(-1)}$ (T3), 14.08 qx.$ha^{(-1)}$ (T2) and 17.49 qx.$ha^{(-1)}$ (T1 full irrigation) (Table 9). In percentage terms, the reduction in grain yield compared with treatment T1 was 74.11% for the rainfed treatment, 53.81% for T3 and 20.77% for T2. Given that the average rainfall during the growing seasons (2022 and 2023) was 98.05 mm, this represented only 39.70% of the average water requirement for the two seasons.

Table 10. Mean squares of ANOVA variances of camelina water use parameters over the two cropping seasons (2021-2022) and (2022-2023).

Source	Treatment	Year	T x A	Groups
Ddl	3	1	3	
EUE grains (Kg.$m^{(-)\ (3)}$ ·ha^{-1})	0.324**	0.15^{NS}	0.20^{NS}	A(T1,T2),B(T3),C(T0)
EUEI gains (Kg.$m^{(-)\ (3)}$ ·ha^{-1})	0.197**	0.336**	0.037^{NS}	A(T1,T2) B(T3)
EUEBio(Kg.$m^{(-)\ (3)}$ ·$ha^{(-1)}$))	0.951*	59.105**	4.874**	A(T2),AB(T1and T0),B(T3)
EUEIBio(Kg.$m^{(-3)}$ ·ha^{-1})	5.100*	0.025^{NS}	$1,569^{NS}$	A(T1 and T2),B(T3)

*WUE: total grain water use efficiency; IWUE: grain irrigation water use efficiency;BioWUE: total biomass water use efficiency;BioWUE: biomass irrigation water use efficiency. freedom;NS:not significant;** significance($p < 0.01$);*significance($p < 0.05$),*

These observations are consistent with other studies that have examined the links between yield, biomass and water efficiency. Zhang et *al.* (2006)) have shown, in the case of wheat, that the highest grain yield was associated with maximum biomass and a high harvest index, and optimum total water use efficiency, but not necessarily with high irrigation water use efficiency. These results highlight the importance of rethinking irrigation management, particularly in semi-arid areas. Indeed, as Rekaby et *al.* point out, (2016)) it is essential, particularly in semi-arid zones, to shift the focus from simple production per unit area to maximising productivity per unit of water consumed, in order to improve the management of water resources.

These results confirm this need, showing that the efficiency of irrigation water use is higher than that of total water use. In particular, T2, with a moderate reduction in irrigation, showed the highest grain and biomass EUEI values (0.96 kg/m^3 and 5.00 kg/m^3 respectively), while T3, with a more severe restriction, showed the lowest values (0.73 kg/m^3 and 3.45 kg/m^3). This clearly illustrates the importance of balanced irrigation management to optimise water productivity without excessively compromising yields, while allowing plants to develop drought resilience in response to progressive levels of water stress. These results are in line with those of Hergert et *al*. (2016)), who found that supplementary water inputs of 210 to 250 mm, i.e. 60 to 70% of full irrigation conditions, produced yields of 78% and 87% of the maximum.

It is also interesting to note that T1 and T2 showed statistically similar efficiencies in terms of total water use, while T1 received 25% more water. This suggests that an increase in water intake does not necessarily translate into a significant improvement in dry matter production, but may, on the contrary, result in additional water loss.

These observations are particularly relevant in the current context of the study region, where water management has become a crucial issue for farmers. Almost all farmers in the Aurès, faced with a gradual decline in crop yields due to water stress, are tending to over-irrigate. Grassini et *al* (2011)) have estimated that 55% of farms irrigate in excess of their water requirements, raising the need to switch to modern irrigation systems and improve irrigation planning and scheduling. Based on tools for monitoring soil moisture and agro-climatic parameters, it is possible to reduce irrigation withdrawals by up to 32%, while minimising yield losses (Grassini et al., 2011) .

The Aurès region is faced with irregular irrigation water pumping due to repeated stoppages caused by the overloading of the electricity network. On a collective level, deficit irrigation appears to be a solution for increasing the efficiency of irrigation water and alleviating the overloading of the electricity network.

In this study, the application of the T2 treatment, aimed at saving 25% of the irrigation water, resulted in an increase in the EUEI of 7.80%. However, this water saving was accompanied by a yield loss of 20.77% on average over the two experimental years. In order to minimise these yield losses under deficit irrigation, farmers need to acquire better knowledge of crop adaptation to water stress and of the combined effect of interannual variability in climatic and soil factors and cropping system management on the WUEI (Geerts & Raes, 2009) . In a very broad context, the aim of this study was to assess the response of various crop-related components under variable agroclimatic conditions over two seasons in a semi-arid environment, under different irrigation treatments. Its aim was to help growers and other stakeholders in the value chain to assess the economic and production consequences of water allocations in regions where water resources are in permanent decline.

2.1.6. Energy and Oil Potential Variables: Oil Yield (RH) and Biodiesel Estimate (EB)

The parameters for estimating biofuels from oilseed crops are of crucial importance for environmental sustainability and modern agriculture. These data make it possible to assess the energy efficiency of crops, promote sustainable use of resources and contribute to the transition towards more environmentally-friendly farming practices. The biofuel estimate and oil yield are closely linked to the oil content and grain yield expressed in litres per hectare (L. ha^{-1}) and kilograms per hectare (Kg. ha^{-1}) respectively. The calculation of oil and biofuel yields showed a similar trend to that of grain yield and oil content in relation to water inputs. Biofuel and oil production increased progressively with increasing water inputs from the different irrigation regimes applied. Furthermore, the significant effect of year and treatment on oil and biofuel yields was demonstrated by an analysis of variance (ANOVA) (Table 11).

Table 11. Mean squares of the ANOVA variances of the Energy and Oil Potential Variables of camelina over the two cropping seasons (2021-2022) and (2022-2023).

Source	Treatment	Year	T x A	Groups
Ddl	3	1	3	
Thuil (%)	35.781**	52.531**	4,781^{NS}	A(T1),AB(T2), BC(T3),C(T0)
RH (Kg. ha^{-1})	177835.957**	173167.891**	6036.091*	A(T1),B(T2),C(T3),D(T0)
EB (L. ha^{-1})	34273.190**	33372.361**	1162.999*	A(T1),B(T2),C(T3),D(T0)

TH : Oil content ;RH ;Oil yield ;EB : Biodiesel estimate ;Ddl :degree of freedom *;NS :not significant ;** significance($p < 0.01$) ;*significance($p < 0.05$),*

Treatment T0 (rainfed treatment) recorded the lowest production values on average over the two seasons, with 101.50 Kg. $ha^{-1}$of oil yield and 44.56 L.$ha^{-1}$of biodiesel (Figure 29). These values increased linearly with the increase in water inputs, reaching the best values corresponding to treatment T1 (full irrigation) with 533.12 Kg. $ha^{-1}$of yield in 2022 and 234.04 L.$ha^{-1}$of biodiesel in 2023, exceeding treatment T2 by 23.3%. In addition, biodiesel production was lower in 2023 than in 2022, which highlights that the differences observed from one year to the next for the main parameters underlines the impact of varying environmental conditions on camelina productivity. (Obour et al., 2017; Wysocki et al., 2013) .

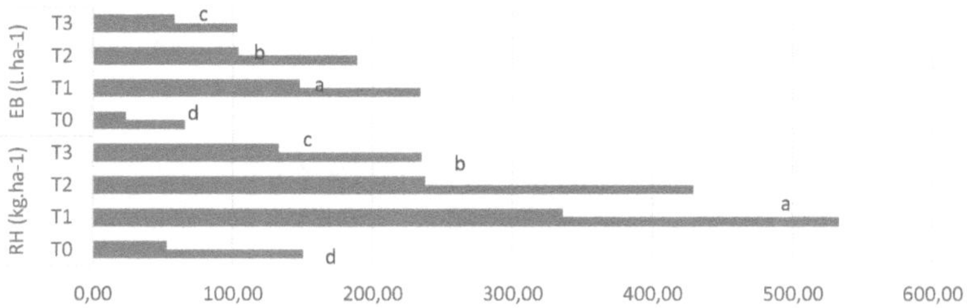

	RH (kg.ha-1)				EB (L.ha-1)			
	T0	T1	T2	T3	T0	T1	T2	T3
2023	52,95	335,83	237,25	132,74	23,25	147,43	104,15	58,27
2022	150,05	533,12	429,22	234,88	65,87	234,04	188,43	103,11

Figure 29. Effect of irrigation regimes on RH and EB during the two seasons (2021-2022) and (2022-2023)

2.1.7. Economic efficiency and electricity parameters

Given the scarcity of water and rising energy prices, the economic efficiency of irrigation strategies is becoming a central issue for sustainable agriculture. However, beyond water management, the energy dimension of irrigation is becoming an increasingly crucial issue for farmers, as it directly affects production costs, particularly in the case of intensive systems such as centralised pivots. Energy consumption by these systems, which are often powered by submersible pumps, accounts for a significant proportion of production costs. In Algeria, where the pricing of electricity subsidised by SONELGAZ (Société nationale de distribution de l'électricité et du gaz) varies according to the hours of consumption, efficient energy management in irrigation is essential to reduce the financial costs associated with irrigation.

The aim of this section is to analyse the economic efficiency and electricity parameters associated with deficit irrigation (T2) and full irrigation (T1) in camelina cultivation under pivots, on an area of 40 hectares. More specifically, we will look at the impact of electricity consumption by pumps and pivots, as well as the potential savings made by reducing the amount of water used, by comparing the two treatments T1 and T2 .

The aim is twofold: 1. to assess the impact of the two treatments on electricity consumption, taking into account the hourly variations in tariffs applied by SONELGAZ.

2- To analyse the trade-offs between the water and electricity savings achieved through deficit irrigation and the loss of grain yield that it causes, and to understand how the choice of irrigation strategies influences not only energy consumption but also overall profitability for farmers.

2.1.7.1. Comparative Energy Analysis of Full and Deficit Irrigation for Camelina under 40 ha Pivot

This work is part of a wider reflection on the sustainability of irrigation systems and their long-term economic impact, both for farmers and for policy-makers.

Let's take the example of the cost of pumping for a pivot irrigation system covering 40 ha, supplied by a borehole equipped with a pump immersed at a depth of 150 metres. The pump, with a flow rate of 56 l/s, ensures that the pivot operates correctly, in accordance with the manufacturer's specifications.

The estimate of the electricity balance is based on the consumption of the submersible pump and the pivot, which operates with 7 electric motors of 0.75 kWh each. According to the operating curves in the pump catalogue, with a flow rate of 56 l/s and a head of 150 m, the power of the pump motor, with an optimum efficiency of over 80%, is 110 kWh, determined from the electrical and dimensional characteristics provided.

The electrical balance is therefore defined as follows :

E balance= E pump+E pivot.

E pump; Pump energy consumption (kWh)

E pivot; The energy consumption of the pivot (kWh).

To draw up this energy balance, it is essential to draw up an irrigation schedule for each treatment for the entire growth cycle of the camelina in order to know the number of hours of operation of the pivot, which was based on the rainfall applied for each irrigation according to the operating catalogue provided by the manufacturer of the ANABIB pivot (Table 12).

The electricity consumption required to irrigate camelina on a 40 ha pivot was calculated for the two treatments: T1 (full irrigation) (Table 13) and T2 (deficit irrigation at 75% ETc) (Table 14). For illustration purposes, we used the irrigation schedule for the second season (2022-2023).

Table 12. Operating data for a 40-hectare irrigation pivot Source: ANABIB manufacturer's operating catalogue.

Water applied (mm)	Timer setting (%)	Ground speed of last reel (m/min)	Elapsed time per lap (hours)
7.84	100	2.49	14.78
8.31	90	2.24	16.42
9.35	80	1.99	18.47
10.69	70	1.74	21.11
11.51	65	1.62	22.74
12.47	60	1.49	24.63
13.60	55	1.37	26.67
14.96	50	1.25	29.56
16.62	45	1.12	32.84
18.7	40	1.00	36.95
21.37	35	0.87	42.22
24.93	30	0.75	49.26
29.92	25	0.62	59.31
37.4	20	0.50	73.89
49.86	15	0.37	98.52
62.33	12	0.30	123.15
83.11	9	0.22	164.21
124.66	6	0.15	246.31
249.32	3	0.07	492.62

Table 13. Electricity Consumption for Complete Irrigation (T1) of the Camelina Cycle under Pivot (40 ha)

Irrigation	Irrigation rate (mm)	Hours of operation	Consumption Pump (kWh)	Consumption Pivot (kWh)	Total
1	11.09	22.74	2501.40	17.055	2518.455
2	13.89	28.00	3080.	21.00	3101
3	9.97	21.00	2310.	15.75	2325.75
4	13.89	25.00	2750	18.75	2768.75
5	18.88	37.00	4070	27.75	4097.75
6	30.69	59.00	6490	44.25	6534.25
7	16.69	32.84	3612.40	24.63	3637.03
8	24.49	49.00	5390	36.75	5426.75
9	13.65	26.67	2933.70	20.0025	2953.7025
10	7.99	16.42	1806.20	12.315	1818.515
Sum	161.20	317.67	34943.70	238.2525	35181.953

Table 14. Electricity consumption for deficit irrigation (T2) of the Camelina cycle under Pivot (40 ha)

Irrigation	Irrigation rate (mm)	Hours of operation	Consumption Pump (kWh)	Consumption Pivot (kWh)	Total (kWh)
1	8.32	16.42	1806.2	12.315	1818.515
2	10.42	21.11	2322.1	15.8325	2337.9325
3	7.48	14.78	1625.8	11.085	1636.885
4	10.42	21.11	2322.1	15.8325	2337.9325
5	14.16	27.67	3043.7	20.7525	3064.4525
6	23.02	45.48	5002.8	34.11	5036.91
7	12.52	24.72	2719.2	18.54	2737.74
8	18.37	38.29	4211.9	28.7175	4240.6175
9	10.24	20.22	2224.2	15.165	2239.365
10	5.99	11.83	1301.3	8.8725	1310.1725
Sum	120.92	241.63	26579.3	181.22	26760.523

2.1.7.2. Economic and Environmental Evaluation of Complete and Deficit Irrigation of Camelina under Pivot on 40 ha

SONELGAZ has adopted an electricity tariff based on hours of consumption as a subsidy for farmers. Electricity consumption during peak hours, between 5 p.m. and 9 p.m., is charged at DZD 8.1147/kWh. For the rest of the day, the rate is 1.7807 DZD/ kWh. This 4.56-fold difference highlights the importance of reducing the duration of irrigation as much as possible during peak hours.

Deficit irrigation (T2) at 75% of crop requirements (ETc) reduced water consumption by 25%, demonstrating increased efficiency in the use of irrigation water. This strategy also resulted in substantial electricity savings of 24%, i.e. a saving of 8421.43 kWh compared with full irrigation during the growth cycle of camelina under pivot irrigation on 40 ha. In particular, electricity consumption during peak hours (from 5pm to 9pm) was reduced by 41% compared with T1, which helps to mitigate the voltage drop on the electricity network during peak hours. This energy saving would enable a medium-sized cold store of 20 $m^{(3)}$ with a power of 5 kW to operate continuously for 70 days.

Satisfying this electricity consumption gap (8421.43 kWh) requires, a natural gas volume of 2105.36 m^3, taking into account a conversion factor according to Gülen, (2011) of 0.25 m^3/kWh and a thermal efficiency of 35% for a gas turbine. This consumption would result in emissions of 4.27 tonnes of CO_2, using the emission factor of 2.03 kg CO_2/m^3, as pointed out by Sterligov, (2020) By adopting T2, it is possible to conserve significant volumes of natural gas, a precious fossil resource, while reducing CO_2 emissions linked to electricity production. Thus contributing to both energy efficiency and environmental protection.

On the other hand, the financial estimate for the two treatments, in accordance with the SONELGAZ scale, showed a financial gain of DZD 26,796 (Tables 15 and 16), which represents a saving of 29.37% compared with the full irrigation treatment (T1). However, the reduction in grain yield associated with the water restriction in treatment T2 resulted in a loss of 117 quintals of camelina, i.e. a financial loss of DZD 819,000 over 40 ha. This evaluation is based on the rapeseed price set at 7,000 DZD per quintal by the Algerian Ministry of Agriculture, due to the absence of a defined price for camelina.

Table 15. Financial estimate of electricity consumption for Full Irrigation (T1) of the Camelina Cycle under Pivot (40 ha)

N Irrigation	Hours of operation		Consumption Pump (Kw)		Consumption Pivot (Kw)		Price before tax (DZD)		Total (DZD)
	From 5pm-9pm	The rest	From 5pm-9pm	The rest	From 5pm-9pm	Off-peak	From 5pm-9pm	Off-peak	
1	2.74	20	301.4	2200	2.055	15	2462	3944	6407
2	4	24	440	2640	3	18	3595	4733	8328
3	1	20	110	2200	0.75	15	899	3944	4843
4	4	21	440	2310	3	15.75	3595	4141	7736
5	4	33	440	3630	3	24.75	3595	6508	10103
6	8	51	880	5610	6	38.25	7190	10058	17247
7	4	28.84	440	3172.4	3	21.63	3595	5688	9282
8	8	41	880	4510	6	30.75	7190	8086	15275
9	4	22.67	440	2493.7	3	17.0025	3595	4471	8066
10	1	15.42	110	1696.2	0.75	11.565	899	3041	3940
Total	40.74	276.93	4481.4	30462.3	30.55	207.69	36613	54614	91227.23

Table 16. Financial estimate of electricity consumption for deficit irrigation (T2) of the Camelina cycle under Pivot (40 ha)

N Irrigation	Hours of operation		Consumption Pump (Kw)		Consumption Pivot (Kw)		Price before tax (DZD)		Total (DZD)
	From 5pm-9pm	The rest	From 5pm-9pm	Off-peak	From 5pm-9pm	Off-peak	From 5pm-9pm	Off-peak	
1	0.00	16.42	0.00	1806.20	0.00	12.32	0	3238	3238.23
2	1.11	20.00	122.10	2200.00	0.83	15.00	998	3944	4941.81
3	0.00	14.78	0.00	1625.80	0.00	11.09	0	2915	2914.80
4	1.11	20.00	122.10	2200.00	0.83	15.00	998	3944	4941.81
5	4.00	23.67	440.00	2603.70	3.00	17.75	3595	4668	8262.83

6	5.48	40.00	602.80	4400.00	4.11	30.00	4925	7889	12813.39
7	4.00	20.72	440.00	2279.20	3.00	15.54	3595	4086	7681.06
8	4.00	34.29	440.00	3771.90	3.00	25.72	3595	6762	10357.23
9	4.00	16.22	440.00	1784.20	3.00	12.17	3595	3199	6793.60
10	0.22	11.61	24.20	1277.10	0.17	8.71	198	2290	2487.35
Total	23.92	217.71	2631.20	23948.10	17.94	163.28	21497	42935	64432.12

2.1.7.3. Cost-Benefit and Environmental Analysis of Complete and Deficit Irrigation

The financial value of the yield loss far exceeds the financial gain in electricity for farmers. This suggests that full irrigation is a more profitable strategy, particularly in regions where water is not a constraint. These findings are supported by the work of Daran et *al*, (2016); Kisekka et *al*, (2016); Payero et *al*, (2008)) who have shown that despite the water savings achieved with deficit irrigation practices, the associated yield losses can lead to a drop in overall profitability . These researchers preferred full irrigation, particularly in regions where water is not a constraint.

However, in semi-arid regions and from the point of view of resource management and sustainability, deficit irrigation presents a considerable advantage for public authorities and decision-makers. The latter are committed to an energy subsidy policy for Algeria's agricultural sector, reaching 65%, which represents an additional burden for the public purse compared with the cost of energy production by gas turbine power stations. These power stations use up fossil natural gas and have a negative environmental impact.

In summary, although deficit irrigation can result in significant water and electricity savings, the associated yield loss and reduced profitability may not be justified for farmers. However, for decision-makers and public authorities, deficit irrigation can be a viable strategy for improving resource management, reducing energy subsidies, promoting the performance of electricity networks and mitigating the negative environmental impacts associated with fossil fuel production.

2.2. Analysis of complex relationships: A multidimensional approach using correlation matrix, PCA analysis and hierarchical tree

2.2.1. Correlation matrix

Exploring the relationships between different variables is a fundamental step in understanding the phenomena under study. Statistical analysis, in particular the use of the correlation matrix, offers a valuable tool for unravelling the complex links between variables and extracting essential information. By quantifying the strength and direction of the relationships between variables, this analysis helps to identify the key factors influencing the phenomenon under study. It helps to understand how the different variables interact, providing a holistic view of the system. It also reveals links that might otherwise go unnoticed by the naked eye, enriching our overall understanding of the subject.

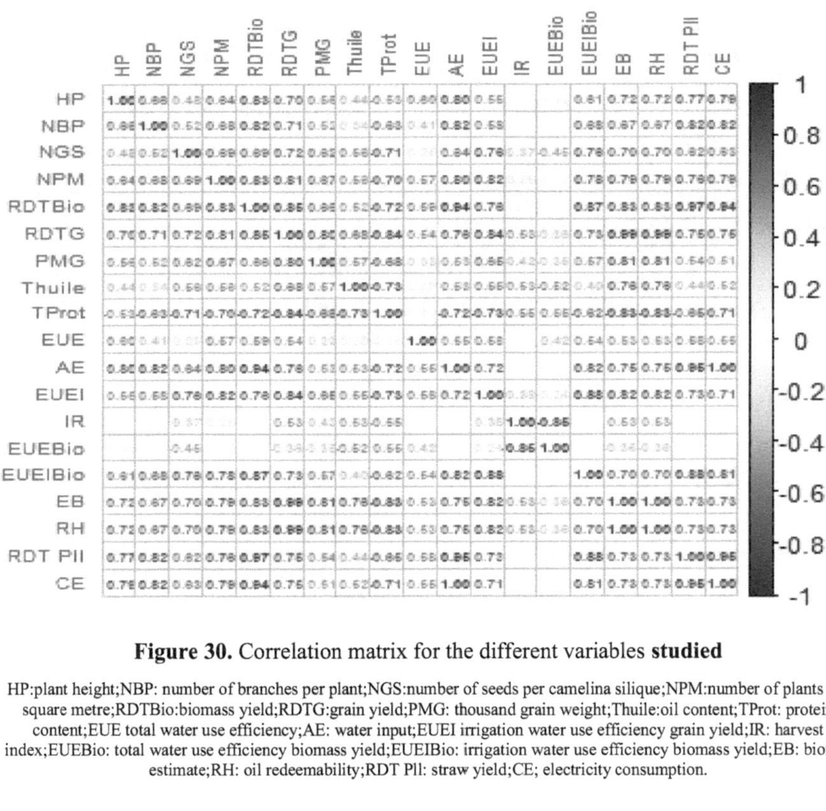

Figure 30. Correlation matrix for the different variables **studied**

HP:plant height;NBP: number of branches per plant;NGS:number of seeds per camelina silique;NPM:number of plants per square metre;RDTBio:biomass yield;RDTG:grain yield;PMG: thousand grain weight;Thuile:oil content;TProt: protein content;EUE total water use efficiency;AE: water input;EUEI irrigation water use efficiency grain yield;IR: harvest index;EUEBio: total water use efficiency biomass yield;EUEIBio: irrigation water use efficiency biomass yield;EB: biofuel estimate;RH: oil redeemability;RDT Pll: straw yield;CE; electricity consumption.

Greater plant height (HP) is associated with more branches per plant (NBP) (r = 0.76**), as well as higher grain yield (RDTG) (r = 0.70**), higher oil content (r = 0.44**) and higher estimated biofuel (EB) (r = 0.72**). Furthermore, these positive relationships increased with increasing water intake (AE) and higher electricity consumption (CE) (r = 0.78**) (Figure 30).

However, these parameters show an inverse trend with protein content (TProt). In fact, protein content increased with water restriction (AE) (r = - 0.72**) and electricity consumption (EC) (r = - 0.71**). A higher protein content is associated with shorter, less branched plants (r = 0.63**) and lower yield productivity (grain or oil) (r = -0.84** and r = -0.83** respectively), as well as lower electricity consumption (EC) (r = - 0.71**) (Figure 30).

Using a simple linear correlation between seed yield and other production parameters ($p \leq 0.05$), the results indicate significant correlations. Seed yield (RDTG) shows a strong positive correlation with plant height (r = 0.70**) and thousand seed weight (MSP, r = 0.80**), highlighting the importance of these factors in improving growth and yield. On the other hand, there was a less significant correlation between seed yield and oil content (r = 0.68*). On the other hand, no significant correlation was observed between seed yield and harvest index (HI). However, it is important to note that a strong negative correlation was reported between seed yield and protein content (r = -0.84**). These results highlight the importance of obtaining better quality yields under deficit irrigation, using less water, with an acceptable loss of yield quantity.

2.2.2. Principal Component Analysis (PCA)

The principal component analysis (PCA) presented below (Figure 31) is a powerful tool for visualising the relationships between variables and simplifying the complexity of the data while retaining their most significant variance, which makes it possible to understand the main axes of variation in the data.

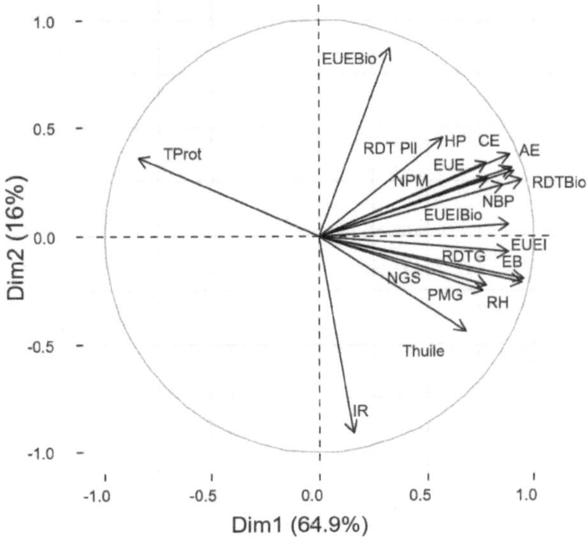

Figure 31. Principal Component Analysis (PCA) of the Relationships between the Variables studied

HP:plant height;NBP: number of branches per plant;NGS:number of seeds per camelina silique;NPM:number of plants per square metre;RDTBio:biomass yield;RDTG:grain yield;PMG: thousand grain weight;Thuile:oil content;TProt: protein content;EUE total water use efficiency;AE: water input;EUEI irrigation water use efficiency grain yield;IR: harvest index;EUEBio: total water use efficiency biomass yield;EUEIBio: irrigation water use efficiency biomass yield;EB: biofuel estimate;RH: oil redeemability;RDT Pll: straw yield;CE; electricity consumption.

The Principal Component Analysis (PCA) plot reveals clear trends among the variables studied. The two dimensions together explain 80.9% of the total variance in the data, with Dim1 representing 64.9% and Dim2 16%. These two axes therefore provide a fairly comprehensive view of the variability of the data.

Dim1 (64.9%): This dimension is strongly correlated with variables such as seed yield (RDTG), plant height (HP), number of branches per plant (NBP), and electricity consumption (CE). A high Dim1 value therefore indicates a tendency to have taller plants, a greater number of branches, a higher seed yield, and higher water and electricity consumption. These variables cluster around the first principal component, indicating a strong correlation between them and suggesting that they share common influences and similar trends.

On the other hand, protein content (TProt) is located at the opposite end of this dimension. This indicates that higher protein content is associated with lower values of the other variables mentioned. This opposite trend with respect to the variables grouped around Dim1 shows that an increase in protein content is associated with a decrease in plant growth and biomass, lower productivity in terms of grain and oil yields, less branched plants, lower water and electricity consumption and a reduced estimate of biofuel production.

Dim2 (16%): The Harvest Index (HI) stands out close to the second principal component, showing its independence from the other variables. It shows unique trends and a low correlation with the variables grouped around the first component. This indicates that the harvest index has distinct characteristics that are not shared with the other variables studied, adding an extra dimension to the analysis.

In summary, the PCA plot highlights the complex relationships between the different variables. While the first dimension captures the main variations associated with plant growth and yield, the second dimension highlights the independence of the harvest index. Protein content shows an inverse trend in relation to growth variables, highlighting the potential trade-offs between yield quality and quantity. These results, in line with the conclusions of the correlation matrix, offer a holistic view of the interactions between the agronomic variables studied.

2.2.3. Hierarchical tree

The hierarchical ascending classification dendrogram is a powerful tool for reinforcing our findings, revealing distinct groupings of variables, suggesting common relationships and influences within each cluster, and providing a clear visual representation of the relationships between variables.

This analysis clearly reveals that our variables have been grouped into six distinct sets of clusters based on their similarities. The variables included in each cluster share common characteristics and show similarities in their behaviours or trends (Figure 32).

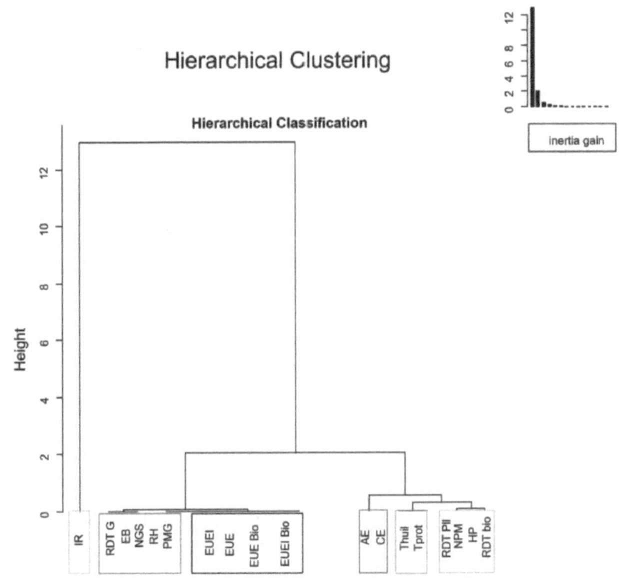

Figure 32. The hierarchical tree.

HP:plant height;NBP: number of branches per plant;NGS:number of seeds per camelina silique;NPM:number of plants per square metre;RDTBio:biomass yield;RDTG:grain yield;PMG: thousand grain weight;Thuile:oil content;TProt: protein content;EUE total water use efficiency;AE: water input;EUEI irrigation water use efficiency grain yield;IR: harvest index;EUEBio: total water use efficiency biomass yield;EUEIBio: irrigation water use efficiency biomass yield;EB: biofuel estimate;RH: oil redeemability;RDT Pll: straw yield;CE; electricity consumption.

Cluster N04 groups together variables related to water use efficiency for grain and biomass yield. This suggests a strong interconnection between these variables, sharing similar trends in response to water availability (Figure 32).

Cluster N05 groups together variables linked to crop productivity such as seed yield (RDTG), thousand seed weight (PMG) and number of seeds per silique (NGS). These variables are highly correlated as they measure different aspects of seed production, as well as oil (RH) and EB biofuel productivity parameters.

Cluster N6, represented by the harvest index (HI), forms its own distinct cluster, suggesting its relative independence from the other variables studied, as indicated by its low correlation with the others.

Cluster N2 groups together two key variables influencing seed quality and composition: oil content (Thuile) and protein content (Tprot). Although these variables show opposite trends, they are strongly negatively correlated with each other.

Cluster N01 groups together variables related to biological yield (RDT bio), plant height (HP) and number of branches per plant (NBP), indicating their interconnection and similar response to growing conditions.

Cluster N3 groups together data on available water intake (AE) and electricity consumption. These variables are closely linked and share a similar trend in response to water restrictions.

Conclusion

In this section of the study, we explored the effects of different irrigation treatments on camelina cultivation during the 2021-2022 and 2022-2023 crop years. The results show that irrigation treatments have a significant impact on water use efficiency for both grain and biomass production. Maximum values of irrigation water use efficiency (EUEI) were obtained under treatment T2, while severe water restrictions under treatment T3 led to minimum values. These observations confirm the hypothesis that moderate levels of water stress can promote more efficient use of water by plants by stimulating resilience mechanisms.

After assessing water use efficiency under different irrigation treatments, it is crucial to understand the overall impact of climate variability on yields and growth parameters. The extreme temperatures and low rainfall observed during the second season (2022-2023) severely affected camelina development and yield. These results highlight the need for optimised irrigation management to maximise camelina productivity in semi-arid environments.

As well as yields in terms of quantity, crop quality is an essential aspect to consider, especially under conditions of water stress. It is also important to stress that water restrictions do not necessarily lead to a proportional drop in yields. In fact, in terms of quality, crops grown under water restrictions have significantly higher protein levels than those grown under full irrigation. This demonstrates that water-stressed conditions can produce higher quality crops, albeit often at the cost of lower overall productivity and a lower oil extraction rate. It is therefore possible to achieve higher quality in terms of protein content despite reduced productivity yields.

In conclusion, the results of this study provide practical guidelines for optimising irrigation management in semi-arid environments. The findings of this part of the study offer valuable information to farmers and decision-makers, enabling them to better understand efficient water management in semi-arid regions for sustainable, high-quality agricultural production while preserving water resources. The renewed interest in *Camelina sativa* in contemporary agriculture marks a promising shift towards sustainable practices. Camelina has a relatively short growth cycle (158 days on average over the two seasons) and low water requirements (246.98 mm on average over the two seasons), making it a viable alternative to fallowing between two cereal crops. This approach makes it possible to avoid monoculture and generate socio-economic benefits, such as improved incomes for farmers.

Growing Camelina under deficit irrigation can make a significant contribution to water conservation and management, making it possible to expand irrigated areas in Algeria using the water saved from deficit irrigation. However, it is important to recognise that the potential seed yield of *Camelina sativa* is still well below that of rapeseed and conventional oilseeds, underlining the need for genetic improvement to increase seed size in order to improve oil extraction conditions and adapt it to the agricultural mechanisation available and dominant in Algeria. This is a possible direction for future research.

We recommend the adoption of treatment T2, which involves continuous deficit irrigation at 75% of ETc (crop water requirements), saving 25% of water with an acceptable yield loss. Although this loss may seem unjustifiable in financial terms for farmers, it represents a viable alternative for decision-makers and public authorities in terms of the sustainable management of water resources in a semi-arid environment, where these resources are in permanent decline. In addition, this strategy would make it possible to reduce the colossal sums allocated to subsidising agricultural electricity by reducing electricity consumption by 24% and by 41% during peak hours, thereby alleviating the overloading of electricity grids while protecting electrical equipment from drilling.

These results are particularly relevant to agricultural strategies in regions faced with limited water resources and high energy costs.

3. Analysis of the Water Efficiency and Fodder Quality of *Camelina sativa* in Response to Deficit Irrigation

In the previous section, we explored the effects of water and climate restrictions on *Camelina sativa* productivity and water use factors. This analysis included a detailed assessment of the 2021-2022 and 2022-2023 cropping seasons, focusing on estimated biofuel, growth, oil, grain, biomass and straw productivity parameters, as well as water use efficiency and economic impact. This research has revealed crucial information on how different climatic conditions and irrigation treatments influence the overall performance of camelina.

Building on this analysis, this section of the study focuses specifically on the impact of deficit irrigation on camelina by-products in the first crop year (2021-2022), namely straw and oilcake, as fodder. While the first part established a solid basis in terms of yield and crop resilience to stress, it is essential to understand how these irrigation practices affect the quality of the by-products used in animal feed.

Introduction

In Algeria, the livestock sector is facing significant challenges due to limited fodder production, exacerbated by unfavourable soil and climate conditions. Soil degradation, lack of rainfall and climatic variability are leading to a reduction in the availability of fodder, which is essential for livestock. This situation means that the country has to rely heavily on fodder imports to meet the growing needs of the livestock sector. Current management of the sector represents a considerable financial burden for public finances, with a direct impact on food security.

Against this backdrop, it is essential to steer fodder production towards new prospects, in particular by introducing new fodder species adapted to different ecological zones. The aim is to diversify fodder resources and find alternatives to make up the shortfall, which seems to be one of the sector's priorities. Modern agriculture favours sustainable practices by integrating new species that consume fewer inputs, combined with water-saving strategies. *Camelina sativa*, an emerging multi-purpose crop with promising qualities, could be the answer to the new directions in fodder production.

Irrigation in semi-arid zones is fundamental to crop production, ensuring optimum plant growth and productivity by compensating for insufficient natural rainfall. However, in many countries, water resources are limited, making it necessary to optimise irrigation management. Deficit irrigation is a water management practice that makes it possible to save this precious resource while maintaining acceptable yield levels. (Howell et al., 2010) . It is crucial to understand how this technique affects not only the main yield but also the quality of by-products such as straw and camelina meal.

A thorough understanding of these effects can guide farmers in managing irrigation to optimise fodder quality, which is essential for animal welfare and economic profitability. Camelina by-products, particularly straw and oilcake, are of great importance as animal fodder. Straw can be used as a coarse supplement, while oilcake, which is rich in protein (Paula et al., 2019) , is used as a high-value feed supplement for livestock. Using these by-products contributes to more sustainable agriculture by reducing waste and maximising the use of available resources.

This study aims to provide valuable information for camel producers and water resource managers, thereby contributing to more sustainable and efficient agriculture. By integrating the results obtained on camel productivity and factors related to water use, this part of the study aims to provide a comprehensive perspective on the management of deficit irrigation. It will thus contribute to more sustainable and resilient agriculture, enabling producers to maximise the value of by-products while preserving water resources. In this way, Algeria could improve its self-sufficiency in fodder and reduce its dependence on imports, thereby strengthening food security and the sustainability of the livestock sector.

Our study aimed to assess the response of camelina straw and meal to limited water inputs, in terms of chemical composition, digestibility and fodder value. In addition, we examined water productivity in forage units in a semi-arid environment in the Batna region.

This section presents and analyses the results obtained concerning the effect of water restriction imposed by the different irrigation treatments adopted. The analyses are focused on several key aspects: the fodder quality of camelina straw and meal, their chemical composition, digestibility, energy value, and water productivity in fodder units. The aim is to understand how deficit irrigation management affects these parameters in a semi-arid environment such as that of the Batna region.

3.1. The effect of irrigation regimes on the chemical composition

Straw

Analyses of the chemical composition of livestock feed are crucial for optimising nutrition, improving animal health and guaranteeing economic efficiency. They help to formulate balanced rations, check feed quality and meet the specific nutritional needs of animals. These analyses are also essential for the development of new feeds. In short, they are essential for ensuring quality feed, animal health and productivity, while optimising resources and costs for farmers.

Table 17. Effect of treatments on the chemical composition of camelina straw

Treatments	MS%	In % of DM				
		MM %	MO%	CB	MAT % (%)
T0	96.17 a ±0.29	6.55 a ±0,06	93.44 c ±0,06	70.18 b ±3,88	4.37 a ±0,22
T1	92,25 c ±0,25	5.30 c ±0,05	94.70 a ±0,05	78.00 a ±2,82	3.55 b ±0,19
T2	92,25 c ±0.32	5.40 c ±0,03	94.59 a ±0,03	76.59 ab ±4,14	3.84 b ±0,12
T3	94.40 b ±0,14	5.78 b ±0.27	94.22 b ±0,27	73.19 ab ±1,66	3.69 b ±0,40

Values followed by the same letter do not show a statistically significant difference at a threshold of $p < 0.05$. DM: dry matter, MM: mineral matter, OM: organic matter, CB: crude fibre, MAT: total nitrogenous matter.

The results show significant variations in the different chemical components of camelina straw depending on the treatment. For example, treatment T0 has the highest dry matter (DM) content, while treatments T1 and T2 show similar and lower values for mineral matter (MM) and dry matter (DM). The differences observed in crude fibre (CF) and total nitrogenous matter (TNM) between treatments suggest potential impacts on digestibility and nutritional value for livestock (Figure 33 and 34).

The results and the ANOVA statistical analysis (Table 17) show a clear increase in OM content in T1 (94.70%) and T2 compared with the rainfed treatment (93.44%) and T3. This suggests that water stress can limit the uptake of nutrients by the plant and the accumulation of organic matter by the plant. The difference in OM between T1 and T2 was small but not statistically significant (p-value > 0.05), indicating that moderate water restriction did not have a major impact on organic matter content. In addition, the results show significant variations in crude cellulose (BC) with a trend similar to that of OM increasing with water comfort. However, the ANOVA shows an opposite trend for MM, MS and MAT, with a significant increase in these components when water restriction intensifies (Akkaya, 1994) . This is explained by the plant's adaptive behaviour under water stress. During periods of water shortage, plants tend to reduce their metabolism and growth, leading to a preferential allocation of available resources towards protein production. As a result, a greater proportion of the plant's resources is used to synthesise proteins, which can lead to an increase in the protein content of plant tissues, including straw. (Richard et al., 2000) .

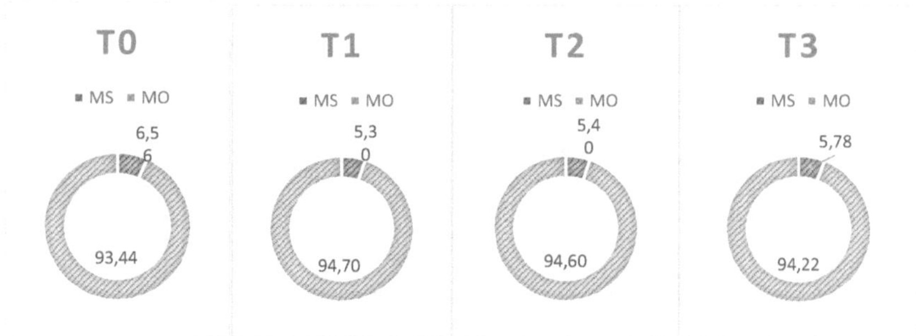

Figure 33. Straw DM and OM for each irrigation regime

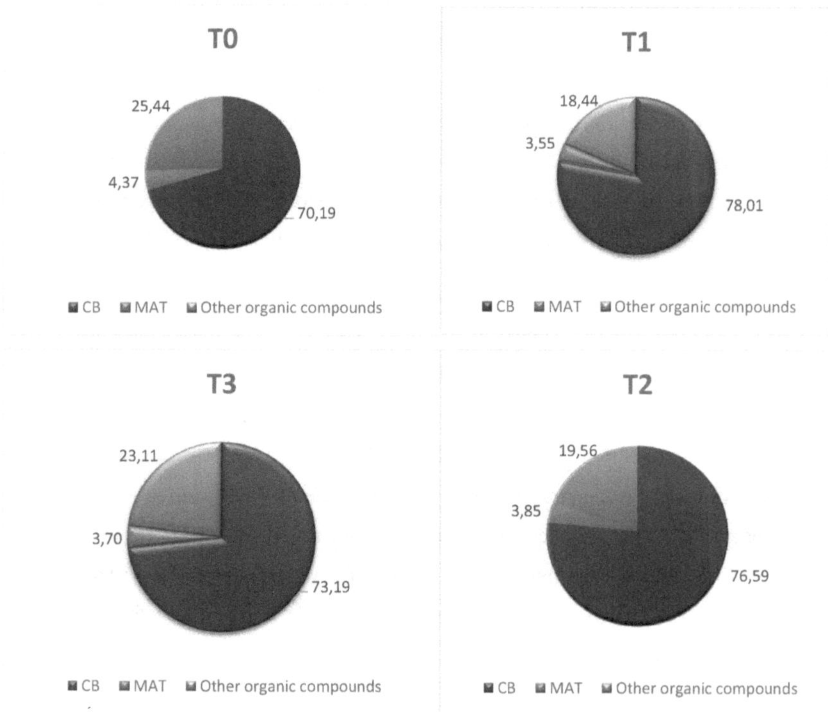

Figure 34. The chemical composition of straw from different treatments in exploded sectors

Oil cakes

Camelina oilcake, a by-product obtained after extraction of the oil from camelina (*Camelina sativa*) seeds, has an interesting chemical composition that makes it versatile in various fields, both food and non-food (Evrard, 2005) . In animal feed, they are used as a nutritional supplement for ruminants and poultry (Juodka et al., 2022) thanks to their high protein and essential fatty acid content, improving the growth and health of animals. In aquaculture, they enrich the diets of farmed fish, promoting their growth and health thanks to their high protein and omega-3 fatty acid content (Hixson et al., 2014) . Camelina oilcake also has non-food applications: in cosmetics, it is used for its moisturising and anti-ageing properties (Delmas et al., 2013) ; in the bioenergy sector, it is used as biomass for the production of biogas or biodiesel (Agarwal et al., 2021) ; in agriculture, they enrich soils with nutrients, improving fertility and water retention (Evrard, 2005) ; and in the chemical industry, they provide raw materials for the production of bioplastics and lubricants (Sydor et al., 2022) . In this way, camelina oilcake adds value to the by-products of oil extraction and supports sustainable and economic practices in various sectors.

The results obtained from analyses of the effect of irrigation regimes on the chemical composition of oilcake are shown in Table 18.

Table 18. Effect of treatments on the chemical composition of camelina oilcake

	MS%	In % of DM			
		MM %	MO%	CB	MAT % (%)
T0	94.00 a ±0.03	7.32 a ±0,06	92.68 c ±0,04	33.33 a ±0.27	40.47 a ±0,02
T1	93.32 b ±0,06	5.86 c ±0,05	94.13 a ±0,05	31.63 d ±0.26	38.94 c ±0,06
T2	93.77 b ±0.06	5.82 c ±0,03	94.18 a ±0,04	32.01 c ±0.54	39.81 b ±0,04
T3	94.04 a ±0,14	6.03 b ±0.27	93.96 b ±0,03	32.99 b ±0.43	40.16 b ±0,04

Values followed by the same letter do not show a statistically significant difference at a threshold of $p < 0.05$. DM: dry matter, MM: mineral matter, OM: organic matter, CB: crude fibre, MAT: total nitrogenous matter.

It seems that the effect of irrigation regimes on the chemical composition of oilcake follows a similar trend to that observed for straw. Dry matter (DM) values varied slightly between treatments, with the highest values observed for T0 (94.00%) and T3 (94.04%), while T1 and T2 showed slightly lower values (93.32% and 93.77% respectively) (Figure 35 and 36). These differences, although minor, are statistically significant, indicating significant variations in dry matter content between treatments.

Mineral matter (MM) content was highest for treatment T0 (7.32%) and lowest for treatments T1 (5.86%) and T2 (5.82%), with T3 having an intermediate value (6.03%). The significant differences between the treatments suggest a variation in the mineral content of the cakes according to the irrigation regimes applied.

Crude fibre (CB) values also show significant differences between treatments, with T0 having the highest value (33.33%), followed by T3 (32.99%), T2 (32.01%), and T1 having the lowest value (31.63%). Similarly, the total nitrogenous matter (TNM) content varied between treatments, with T0 having the highest value (40.47%), followed by T3 (40.16%) and T2 (39.81%). T1 had the lowest value (38.94%), with all these differences being statistically significant ($P<0.05$). In contrast, organic matter (OM) was highest for treatments T1 (94.13%) and T2 (94.18%), closely followed by T3 (93.96%). Treatment T0 had the lowest value (92.68%). These differences are also significant, indicating a significant influence of the irrigation treatments on the organic matter content of the oilcake.

The results suggest an increase in organic matter (OM) content for treatments receiving more irrigation (T1 and T2), while crude fibre (CF), mineral matter (MM), dry matter (DM) and total nitrogen matter (TNM) appear to increase with a reduction in water supply, indicating a higher concentration of these protein components in oilcake produced under water stress conditions. These observations corroborate the results of Macedo et *al* (2016)), reinforcing our conclusions. This trend can be explained by the fact that, under water stress conditions, plants can allocate more resources to protein production as a means of adapting to their environment. This results in an increase in protein content in crops grown under water stress, as suggested by Cheng et *al.* (2016)).

In summary, irrigation regimes have a significant impact on the chemical composition of camelina oilcake, with potential implications for its nutritional value and use in animal feed.

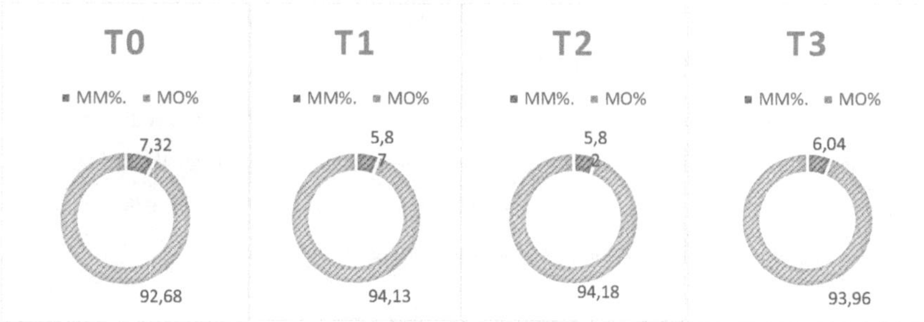

Figure 35. DM and OM of oilcake for each irrigation regime

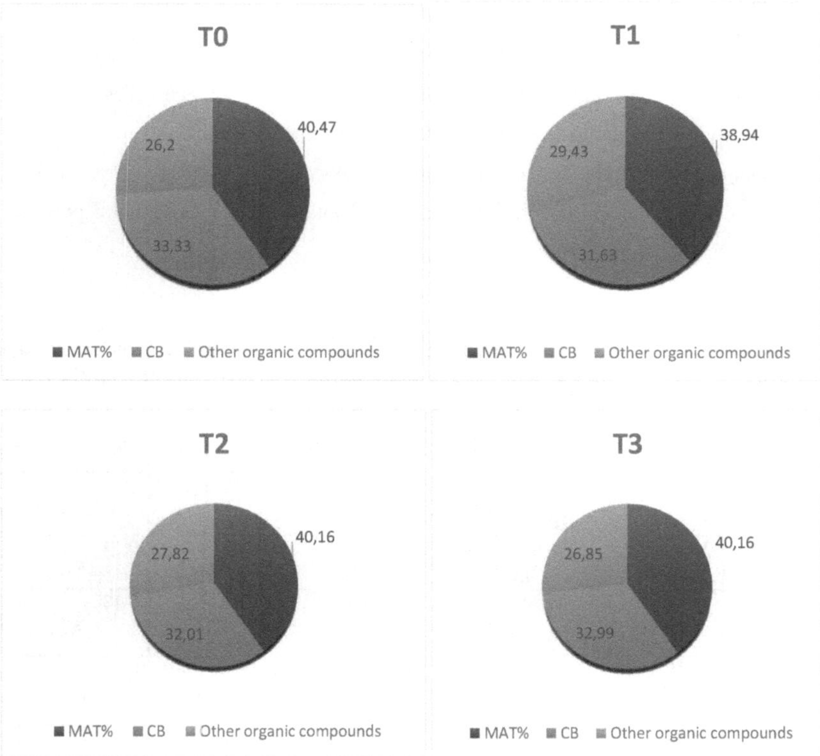

Figure 36. Chemical composition of oilcake from different treatments in split sectors

3.2. The effect of irrigation regimes on enzymatic digestibility and energy value.

After analysing and interpreting the results of the chemical composition of camelina straw and oilcake, it is crucial to assess their feed quality in greater depth in terms of enzymatic digestibility and energy value. These two parameters are essential for determining the nutritional value and utilisation efficiency of the feed for the animal.

Enzymatic digestibility measures the proportion of dry matter that is broken down by digestive enzymes, providing a direct indication of the nutritional quality of the forage. High digestibility suggests that the nutrients in the feed are readily available to animals (Colburn et al., 1968) , which is essential for their production, growth and well-being.

Energy value is a measure of the energy that animals can obtain from feed after digestion. It is generally expressed in terms of Feed Units (FU) per kilogram of dry matter (Shang, 2015)

.

In this section, we will analyse the results of the enzymatic digestibility and energy value of camelina straw and meal samples subjected to different irrigation treatments. These analyses will enable us to gain a better understanding of the impact of irrigation regimes not only on the chemical composition, but also on the nutritional and energy quality of forages. The results of these analyses will provide valuable information for optimising the agricultural practices adopted in the field of animal feed,

Straw

The results of the analysis of the enzymatic digestibility and energy value of the straw from each treatment are shown in Table 19.

Table 19. Effect of treatments on enzymatic digestibility and energy value of camelina straw

Treatments	In % of DM		MOD (%)	WORLD (%)	VE (UF/KgMS)
	DcellMO (%)	BMD (%)			
T0	25.27 a ±0,47	40.62 a ±0,41	38.20 a ±3,14	55.22 b ±2.49	0.14 a ±0,01
T1	24.84 a ±0,63	40.90 a ±0,30	38.40 a ±3.78	56.20 a ±4.24	0.13 a ±0,01
T2	25.20 a ±0,63	40.85 a ±0,41	38.60 a ±4,06	55.90 ab ±3.81	0.14 a ±0,01
T3	24.94 a ±0.80	40.68 a ±0,53	38.30 a ±4.61	55.80 ab ±5.90	0.13 a ±0,01

Values followed by the same letter do not show a statistically significant difference at a threshold of $p < 0.05$. DcellMO: cellulase digestibility of organic matter, DMO: digestibility of organic matter, MOD: digestible organic matter, MOND: digestible organic matter, VE: energy value.

The results of the analysis of the enzymatic digestibility and energy value of the straw show minor variations between the irrigation treatments, with no significant differences ($P<0.05$) for most parameters. The values for cellulose digestibility in organic matter (DcellMO) oscillated between 24.84% (T1) and 25.27% (T0), indicating that the irrigation regimes had no major impact on this parameter. Similarly, organic matter digestibility (OMD) remained relatively stable between 40.62% (T0) and 40.90% (T1). Digestible organic matter (DOM) values varied little, from 38.20% (T0) to 38.60% (T2). Non-digestible organic matter (NDOM) showed a slight variation, with the highest value for T1 (56.20%) and the lowest for T0 (55.22%), with treatments T2 and T3 showing intermediate values (55.90% and 55.80%) respectively, the differences being significant between T1 and T0. Finally, the energy values in fodder units per kilogram of dry matter (FU/KgDM) were virtually constant from 0.13 (T1, T3) to 0.14 (T0, T2), with no significant differences (Figure 37).

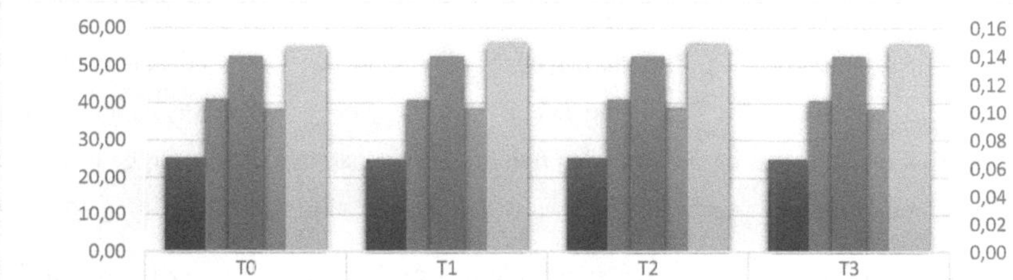

	T0	T1	T2	T3
DcellMO	25,28	24,84	25,20	24,94
DMO	40,90	40,62	40,86	40,69
MOD	38,22	38,46	38,64	38,33
MOND	55,22	56,24	55,95	55,89
UF	0,14	0,14	0,14	0,14

Figure 37: Enzymatic digestibility parameters and energy value of straw from different treatments

These results suggest that the different irrigation regimes do not significantly affect the enzymatic digestibility and energy value of camelina straw, with the exception of non-digestible organic matter (NDI), which was influenced by the treatments applied. This indicates that the parameters (DcellMO, DMO, MOD, and UF) react in a relatively stable manner and are not strongly influenced by the irrigation regime. The slight variations observed can be attributed to the plant's adaptive response to water availability.

Oil cakes

The results of the analysis of the enzymatic digestibility and energy value of the straw from each treatment are shown in Table 20.

Table 20. Effect of treatments on enzymatic digestibility and energy value of camelina oilcake

Treatments	In % of DM		MOD (%)	WORLD (%)	VE (UF/KgMS)
	DcellMO (%)	BMD (%)			
T0	65.16 b ±1.07	69.68 d ±0,04	66.61 b ±0.59	27.89 c ±0.02	0.72 c ±0,03
T1	63.54 c ±0,08	72.16 a ±0,02	65.64 c ±0.06	28.47 a ±0.03	0.77 a ±0,03
T2	63.40 c ±0,02	70.36 b ±0,04	64.54 d ±0.02	28.12 b ±0.02	0.74 b ±0,03
T3	67.40 a ±0.03	69.75 c ±0,02	67.80 a ±0.02	26.16 d ±0.02	0.71 c ±0,03

Values followed by the same letter do not show a statistically significant difference at a threshold of $p < 0.05$.
DcellMO: cellulase digestibility of organic matter, DMO: digestibility of organic matter, MOD: digestible organic matter, MOND: digestible organic matter, VE: energy value.

Irrigation regimes had a significant effect on the enzymatic digestibility of camelina oilcake, as well as on its energy value. Unlike the parameters of enzymatic digestibility of straw, where only MOND was affected by the irrigation regimes, all the parameters seem to be influenced by the different irrigation regimes in the case of camelina oilcake.

The results of the analysis of the enzymatic digestibility and energy value of camelina oilcake reveal significant variations between treatments. The digestibility of cellulose in organic matter (DcellMO) varied from 63.40% (T2) to 67.40% (T3), with T0 and T1 showing 65.16% and 63.54% respectively, indicating a significantly higher digestibility for T3. Organic matter digestibility (OMD) was highest for T1 (72.16%), closely followed by T2 (70.36%), T3 (69.75%) and T0 (69.68%), with these differences being statistically significant. For digestible organic matter (DOM), T3 had the highest value (67.80%), while T2 had the lowest (64.54%), with T0 and T1 showing intermediate values of 66.61% and 65.64%. For non-digestible organic matter (NDM), values ranged from 26.16% (T3) to 28.47% (T1), with T0 and T2 showing 27.89% and 28.12% respectively, indicating that T3 had the lowest NDM content and T1 the highest (Figure 38). The results indicate that treatments with abundant irrigation (T1 and T2) increase the parameters of organic matter digestibility (OMD), non-digestible organic matter (NDOM), and energy value (UF). On the other hand, the water restriction conditions favoured an increase in the other enzymatic digestibility variables, namely the digestibility of cellulose as organic matter (DcellMO) and digestible organic matter (MOD).

DMO, MOND and energy value (UF) showed a positive correlation of 0.86**, 0.61* and 0.86** respectively with water inputs, with maximum values corresponding to treatment T1, which offered full irrigation. On the other hand, minimum values are observed for treatment T3, which offers reduced irrigation. These results suggest that higher levels of irrigation favour better enzymatic digestibility and a higher energy value of camelina oilcake.

On the other hand, cellulase digestibility of organic matter (DcellMO) and (MOD) show a negative correlation -0.61* and -0.60*respectively with water inputs. The maximum values for these parameters are observed for treatment T3, characterised by reduced irrigation, while the minimum values are observed for treatment T2, which offers irrigation at 75% of water requirements. This suggests that lower levels of irrigation may favour better DcellMO and more digestible organic matter in camelina oilcake.

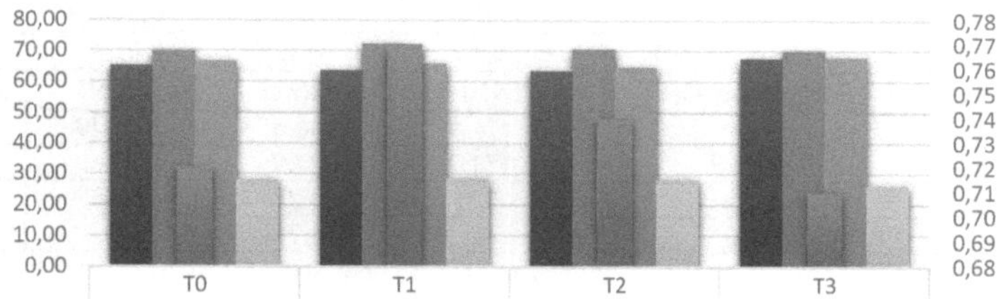

	T0	T1	T2	T3
Dcell MO (%)	65,16	63,54	63,40	67,40
BMD (%)	69,68	72,16	70,36	69,75
MOD (%)	66,61	65,64	64,54	67,80
MON (%)	27,90	28,47	28,13	26,16
VE (kg/MS)	0,72	0,77	0,74	0,71

Dcell MO (%) BMD (%) MOD (%) MON (%) VE (kg/MS)

Figure 38. Enzyme digestibility parameters and energy value of oilcake from different treatments

3.3 The effect of irrigation regimes on water productivity per forage unit.

Having completed the analysis and interpretation of the results of the effects of limited water inputs on the chemical composition, enzymatic digestibility and energy value of camelina by-products, we will finally examine water productivity per forage unit. This is a key parameter that is crucial for water management in animal nutrition.

Water productivity per forage unit (WUP) is an essential parameter for assessing the effectiveness of irrigation strategies and water management in agriculture. It represents the quantity of fodder produced per unit of water used, measured in terms of energy value (UF) per millimetre of water consumed. This parameter is crucial, particularly in regions where water resources are limited, as it enables fodder production to be maximised while optimising water use. Improved water productivity means that crops can provide high nutritional value with minimal use of this precious resource, thus contributing to the sustainability and profitability of livestock production systems. By assessing water productivity per forage unit for camelina by-products under different irrigation regimes, we can identify the most effective water management practices to improve forage production while conserving water resources.

Table 21. Calculation of water productivity per forage unit for straw .

Treatments	T0	T1	T2	T3
RDT Straw (Kg /Ha)	2537.50	7722.50	6790	3555.00
MS (%)	96.17	92.25	92.25	94.40
RTD (Kg DM/ha)	2440.31	7124.01	6263.78	3355.92
Feed value (UF/Kg DM)	0.14	0.13	0.14	0.13
Contribution per hectare	341.64	926.12	876.93	436.27
Input (mm)	-	136.65	102.49	47.83
ET (mm)	151.2	287.85	253.69	199.03
PEF (UF/mm)	2.26	3.22	3.46	2.19
PEIF (UF/mm)	-	4.26	5.22	1.98

Table 22: Calculation of water productivity per feed unit for oil cakes

Treatments	T0	T1	T2	T3
RDT Grain (Kg / Ha)	662.50	2077.50	1785.00	1045
Oilcake (%)	72.00	68.00	70.25	72.25
RDT Oil cake (Kg /Ha)	477.00	1417.89	1253.96	755.01
MS (%)	94.00	93.32	93.77	94.04
Oil cake yield (KgMS/Ha)	448.38	1323.18	1175.84	710.01
Feed value (UF/KgMS)	0.72	0.77	0.74	0.71
Contribution per hectare	322.83	1018.85	870.12	504.11
Input (mm)	-	136.65	102.49	47.83
Water Total (mm)	151.20	287.85	253.69	199.03
PEF (UF/mm)	2.14	3.54	3.43	2.53
PEIF(UF/mm)	-	5.09	5.34	3.79

The results of the tables reveal a notable variation in water productivity in forage units per millimetre of water consumed (FU/mm) between the different irrigation treatments for straw and oilcake. Treatment T1 for oilcake had a high total water productivity (TWP) value (3.54

FU/mm), followed closely by T2 (3.43 FU/mm) (Table 22). Similarly, T2 for straw stands out as having the highest TWP (3.46 UF/mm), slightly outperforming T1 (full irrigation) which has a productivity of 3.22 UF/mm (Table 21). This disparity is even more evident when we examine the irrigation water productivity (IWP) for straw and oilcake, where treatment T2 (straw) clearly stands out with 5.22 UF/mm compared with T1 (4.28 UF/mm), recording an increase of 18%. The disparity is less pronounced for oilcake, with a 4.68% increase in PEIF for T2, with 5.09 UF/mm for T1 and 5.34 UF/mm for T2.

The results highlight the efficiency of the deficit irrigation of treatment T2 in the production of forage units per millimetre of irrigation water applied, compared with treatment T1 with full irrigation. Although T1 received 25% more water than T2, an increased amount of water can lead to additional water losses without significantly increasing the energy value. This finding thus highlights the superiority of the water use efficiency of treatment T2 in the forage production process per millimetre of irrigation water applied, compared to full irrigation T1.

In summary, these results indicate that T2, despite a 25% reduction in irrigation, manages to optimise the use of water to produce forage units, demonstrating a more efficient and sustainable approach to growing camelina under limited water conditions.

3.4. Analysis of complex relationships: correlation matrix, principal component analysis and dendrogram of the parameters studied

In this section, we examine the complex relationships between the different variables studied using advanced statistical techniques. The correlation matrix was used to identify linear relationships between variables. Next, a principal component analysis (PCA) was performed to reduce the dimensionality of the data while retaining as much of the original variance as possible. Finally, a dendrogram was constructed to visualise the hierarchical relationships between the different parameters.

These methods provide an overview of the interrelationships between the parameters, making it possible to identify groupings of variables and understand the underlying factors influencing the chemical composition, enzymatic digestibility and energy value of camelina by-products. This in-depth analysis will provide additional insights and contribute to the overall understanding of the effects of irrigation treatments on camelina by-products.

3.4.1. Correlation matrix

The correlation matrix quantifies the strength and direction of the linear relationships between the different variables.

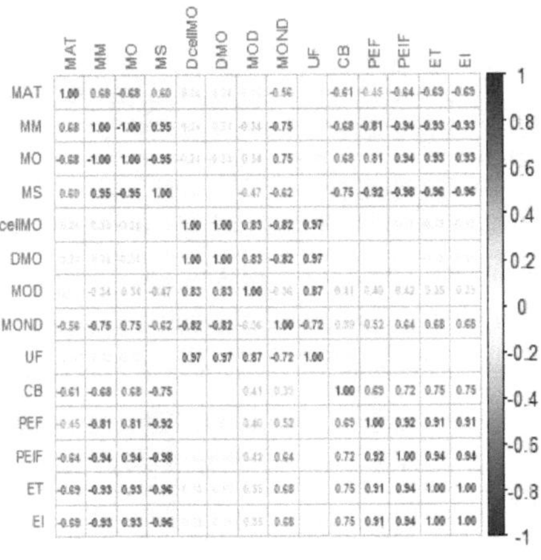

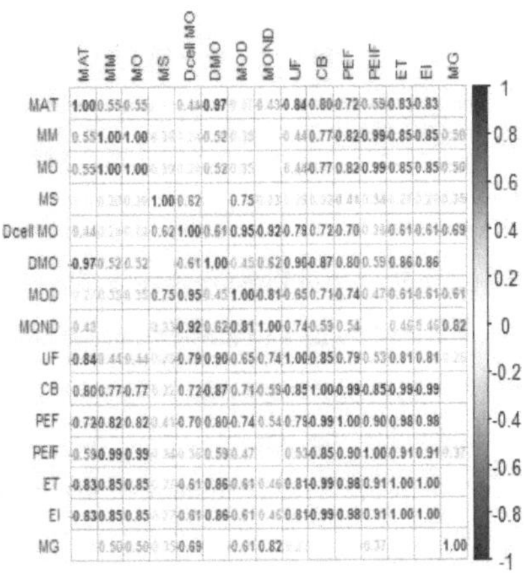

Figure 39A: Correlation matrix Straw **Figure 39B:** Cake correlation matrix

DM: dry matter, MM: mineral matter, OM: organic matter, CB: crude cellulose, MAT: total nitrogenous matter, DcellMO: cellulase digestibility of organic matter, DMO: digestibility of organic matter, MOD: digestible organic matter, MOND: digestible organic matter, VE: energy value, PEF: total water productivity per forage unit, PEIF: irrigation water productivity per forage unit; ET: total water, EI: irrigation water.

Interpretation of the correlation matrices for camelina straw and oilcake reveals some interesting results. In the case of oilcake, a strong positive correlation was observed between VE and water intake (r=0.81**), highlighting the importance of this factor in improving the energy value of oilcake. On the other hand, it is relevant to note that a significant negative correlation was observed between the energy value (UF) on the one hand and (CB) and (MAT) on the other, with correlation coefficients of -0.84** for both, suggesting that better quality yields can be obtained with the use of deficit irrigation (Figure 39B).

On the other hand, for straw, UF showed weaker correlations with the other chemical composition variables. This can be explained and confirmed by the results of the ANOVA, which show an absence of significant variation in the energy value, whatever the irrigation regime applied (Fig. 39A).

3.4.2. Principal Component Analysis (PCA)

PCA is a dimensionality reduction technique that transforms the original variables into a new set of uncorrelated variables called principal components. These principal components explain most of the variance in the data, making it possible to visualise the complex relationships between variables in a reduced-dimensional space.

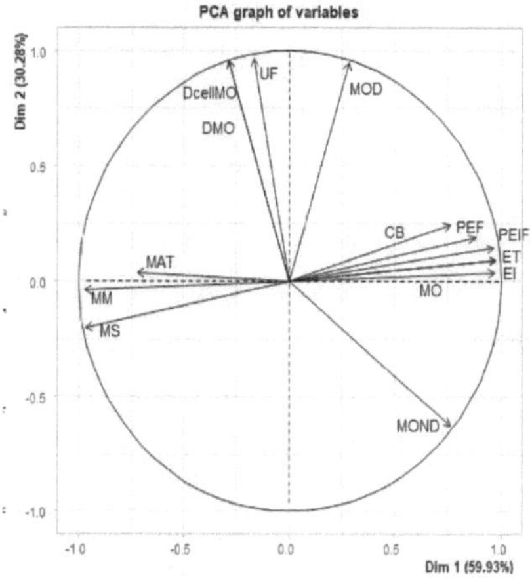

Fig.40A: PCA graphic presentation of the variables studied straw

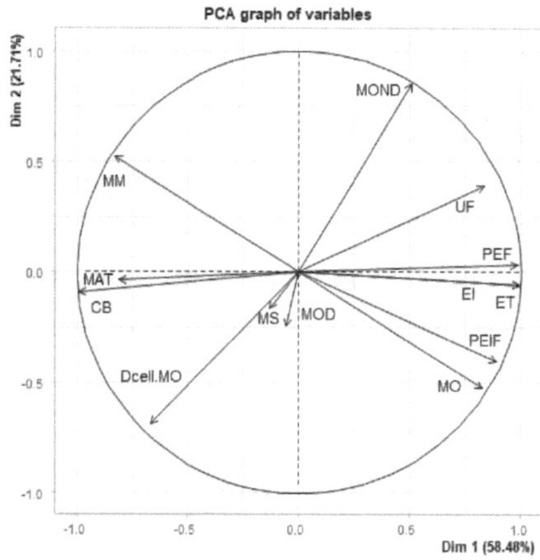

Figure 40B: Graphical presentation of the PCA for the variables studied oilcake

DM: dry matter, MM: mineral matter, OM: organic matter, CB: crude cellulose, MAT: total nitrogenous matter, DcellMO: cellulase digestibility of organic matter, DMO: digestibility of organic matter, MOD: digestible organic matter, MOND: digestible organic matter, VE: energy value, PEF: total water productivity per forage unit, PEIF: irrigation water productivity per forage unit; ET: total water, EI: irrigation water.

Principal component analysis (PCA) offers an additional and complementary perspective to the previous statistical analyses. For the straw PCA, parameters such as (OM), (BC) and water inputs move in the same direction and are grouped close together on the first principal component. This proximity suggests a strong correlation between these variables, indicating that they share common influences and trends (Figure 40A). On the other hand, on the other side of the axis of the first principal component, parameters with an inverse trend stand out, such as (MM), (MAT) and (MS). These parameters appear to be moving in a different direction to those grouped near the first principal component. As for the parameters of enzymatic digestibility and energy value of straw, they stand out near the second principal component, which suggests that they are independent of the other parameters studied.

On the other hand, the PCA of the oilcake results shows a clear dispersion of the variables on the two principal component axes (Figure 40B), with an opposite trend between the parameters MAT, MS and MM on the one hand, and UF, MO, PEF and PEIF on the other, due to the behaviour of these parameters in relation to irrigation water inputs (EI).

3.4.3. The hierarchical tree (Dendrogram)

The dendrogram is a hierarchical classification method that represents the relationships between variables in the form of a tree, making it possible to visualise and identify groups of variables with similar behaviour, providing an intuitive understanding of the complex interrelationships between the parameters under study.

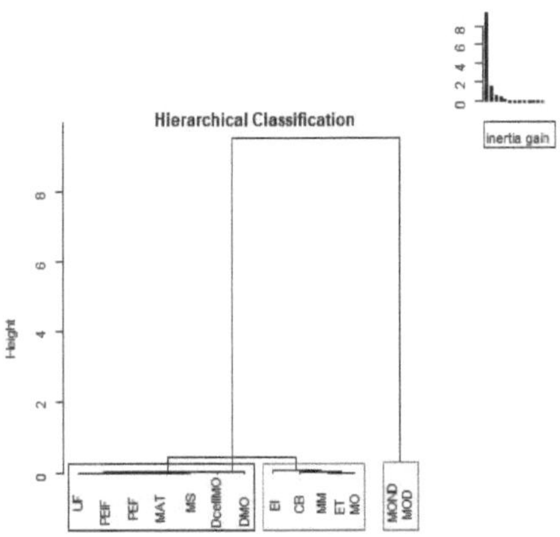

Figure 40A: Hierarchical tree of variables studied straw

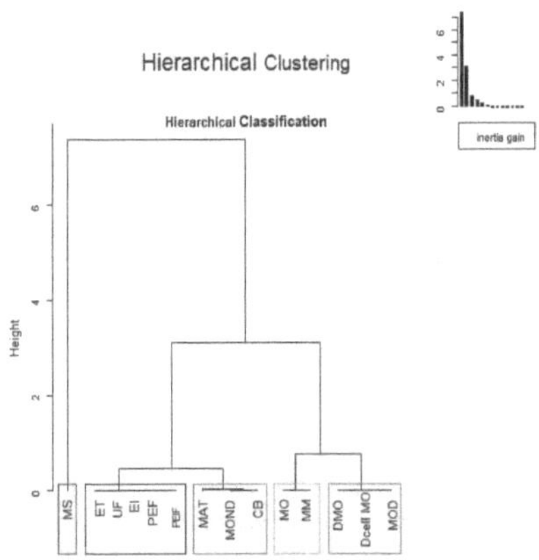

Figure 40B: Hierarchical tree of variables studied cakes

DM: dry matter, MM: mineral matter, OM: organic matter, CB: crude cellulose, MAT: total nitrogenous matter, DcellMO: cellulase digestibility of organic matter, DMO: digestibility of organic matter, MOD: digestible organic matter, MOND: digestible organic matter, VE: energy value, PEF: total water productivity per forage unit, PEIF: irrigation water productivity per forage unit; ET: total water, EI: irrigation water.

Hierarchical tree analysis is proving to be a robust tool for consolidating our earlier findings by grouping variables into clusters based on their similar characteristics. This graphical representation makes it possible to clearly distinguish groupings based on similarities between variables, where the parameters included in each group share common characteristics as well as similar trends in their behaviour.

The (UF) of the oilcakes are grouped with the water parameters, such as total water (ET) and irrigation water (EI), highlighting the significant impact of the water regimes on the energy value of the UF oilcakes, thus linking them closely to the water inputs (Fig.40B) On the other hand, the UF in the straw results were found to be associated with enzymatic digestibility variables, such as (DcellMO) and (DMO), but not with the water variables ET and EI. This dissociation suggests that the effects of water inputs on straw UFs are not significant. Furthermore, the water variables ET and EI are grouped with organic matter (OM), mineral matter (MM) and crude fibre (CB), parameters that are themselves strongly influenced by water regimes (Figure 40A).

Conclusion

This discussion highlights the significant impact of deficit irrigation on the chemical composition, enzymatic digestibility and fodder value of camelina straw and meal. The variations observed in the various forage quality parameters reflect the adaptive responses of camelina to the water stress conditions induced by the irrigation regimes. This highlights the camel's ability to adjust its physiology to cope with changing environmental conditions.

The results underline the crucial importance of water management in the forage production process and highlight the choice of T2 treatment rather than T1. This choice would enable farmers to save up to 25% water while maintaining a quality of straw and cake favourable to animal feed, characterised by a higher protein content than that obtained with a full irrigation regime.

This approach has a number of significant advantages:

Optimisation of water resources: Allows more efficient use of this precious resource.

Environmental sustainability: Helps preserve water resources, which is crucial in regions prone to drought or water scarcity.

Economic profitability: Reducing water consumption translates into cost savings for farmers, by cutting irrigation-related expenses while maintaining satisfactory crop yields.

The absence of any significant difference between treatments T1 and T2 in terms of several parameters indicates that it is possible to reconcile irrigation efficiency, environmental sustainability and economic profitability by adopting deficit irrigation practices on Camelina crops. This is particularly relevant in the context of the need to strike a balance between water availability, crop yields and the search for reliable alternative solutions for animal feed in semi-arid regions.

The results of our study are particularly important for the introduction of new drought-resistant species, especially in vulnerable areas where farmers have to make strategic decisions about crops and irrigation practices. These decisions are crucial for maximising farm and livestock profitability while preserving the sustainability of agriculture. Camelina could thus become a major alternative fodder crop for agriculture in the semi-arid zone of North Africa, offering favourable prospects for improving and diversifying the fodder supply system in Algeria.

It is essential to continue in-depth research on camelina in the North African region in order to fully exploit its characteristics and make an effective contribution to the resilience of local farming systems. This approach also highlights the importance of continuing to study the physiological behaviour of camelina in order to better control the choice of the right phenological stage for its use as fodder.

General conclusion

The study highlighted the crucial importance of water management in Algerian agriculture, particularly in semi-arid areas where water resources are increasingly limited. The results of this study provide valuable knowledge for farmers and decision-makers, enabling them to better understand efficient water management in semi-arid regions for sustainable, high-quality agricultural production while preserving water resources. By examining the cultivation of camelina (*Camelina sativa*) under different irrigation regimes, we were able to demonstrate that deficit irrigation, in particular the T2 treatment, offers a viable compromise between saving water and reducing yield.

The reappearance of camelina sativa in contemporary agriculture marks a promising shift towards sustainable practices. Camelina has a short growing cycle (158 days) and low water requirements (246.98 mm), making it a viable alternative to fallowing between two cereal crops. This approach makes it possible to avoid monoculture and replace failed crops in the event of early losses of winter crops, with socio-economic benefits such as improved financial income for farmers.

The results obtained show that camelina is much more promising for farming systems in marginal areas when combined with the deficit irrigation strategy. In particular, the T2 treatment, at 75% of water requirements, enabled significant water savings while maintaining acceptable yield levels. This highlights the potential effectiveness of deficit irrigation in improving water use efficiency (WUE), while mitigating the negative impacts of water stress on agricultural production.

In addition, the T2 treatment demonstrated greater efficiency in the use of irrigation water. It also resulted in substantial electricity savings of 24%, i.e. a saving of 8421.43 kWh during the growth cycle of camelina under pivot irrigation on 40 ha. This energy saving is equivalent to 70 days of continuous operation of a medium-sized cold room with a capacity of 20 m^3 and a power of 5 kW. In particular, there was a significant 41% reduction in electricity consumption at peak times compared with full T1 irrigation, which helps to reduce the voltage drop on the electricity network.

The production of 8421.43 additional kWh requires the extraction of 2105.36 m^3 of natural gas, resulting in the emission of 4.27 tonnes of CO_2 . By adopting T2, it is possible to significantly limit the extraction of this fossil resource and reduce the greenhouse gas emissions associated with electricity generation. This strategy therefore represents a significant step forward in terms of energy efficiency and ecological transition.

The benefits of T2 treatment don't stop there - crops grown under moderate T2 water restrictions have higher protein levels than those grown under full irrigation, although this often results in lower overall productivity and a lower oil extraction rate.

On the other hand, the severe water restrictions represented by treatment T3 (35% of water requirements) resulted in minimum EUEI values associated with a serious alteration in the main components of crop productivity and growth. The adverse consequences of this severe restriction were evident in the second crop year (2022-2023), where the extreme temperatures and low rainfall observed remarkably affected camelina development and yield. These observations confirm the hypothesis that moderate levels of water stress can encourage plants to use water more efficiently by stimulating resilience mechanisms.

In terms of agricultural policies, this study suggests adopting sustainable irrigation strategies and improving irrigation planning, based on accurate determination of water requirements using lysimeters. These measures could lead to a significant reduction in water withdrawals while maintaining agricultural productivity.

The co-product valorisation section highlighted the significant impact of deficit irrigation on the chemical composition, enzymatic digestibility and fodder value of camelina straw and oilcake, as well as on the water productivity per fodder unit resulting from each treatment. The results demonstrate the crucial importance of water management in the forage production process. In particular, treatment T2, compared with T1, enables farmers to save up to 25% water while maintaining a favourable quality of straw and cake for animal feed. The straw and oilcake from T2 have a higher protein content than those from T1, and the T2 treatment is also clearly superior in terms of water productivity per forage unit, underlining its efficiency. This approach strengthens the Water-Food Security nexus, in a context of optimising water resources, environmental sustainability and economic profitability for farmers.

The results of this study are of considerable importance for the introduction of new drought-resistant species, particularly in vulnerable areas where farmers have to make strategic decisions about crops and irrigation practices in order to maximise the profitability of agriculture and livestock while preserving agricultural sustainability. Camelina could indeed become an important alternative fodder crop for agriculture in the semi-arid zones of North Africa, offering promising prospects for improving and diversifying the food system in Algeria.

Outlook:

Growing *Camelina sativa* is not only a viable solution for diversifying agriculture in Algeria, but also a model for more sustainable farming practices that are resilient in the face of climate change. Integrating this crop into farming systems could help to make better use of limited water resources, while offering substantial economic and environmental benefits.

Future research should focus on genetic improvement of camelina to increase yields by increasing seed size and eliminating constraints linked to mechanisation and processor concerns. At the same time, it is crucial to optimise irrigation techniques to maximise water use efficiency. It is also necessary to modernise and extend the coverage of the network of meteorological stations in Algeria in order to accurately determine irrigation doses and avoid any form of over-irrigation.

In addition, it is crucial to raise awareness and train farmers in techniques for calculating water requirements and the disastrous consequences of wasting water and the gradual decline in groundwater.

However, further in-depth research into camelina in the North African region is imperative in order to fully exploit its intrinsic characteristics and facilitate its significant contribution to building the resilience of local farming systems. This concerted effort is essential to improve its competitiveness in relation to other prevalent oilseed crops, facilitating its transition from a niche crop to a mainstream agricultural commodity . This approach highlights the importance of continuing to study the physiological behaviour of camelina in order to gain a better understanding of the choice of the right phenological stage for its use as fodder.

References :

Aboukhaled, A., Alfaro, J. F., & Smith, M. (1982). *Lysimeters. United Nations Food and Agriculture Organization, Irrigation and Drainage Paper 39, Rome, 68 p. https://books.*google.dz/books?hl=en&lr=&id=OGTieeN5tT0C&oi=fnd&pg=PA9&dq=bulletin+FAO+39+Lysimeters&ots=r8hhNQfVZV&sig=R7gXsY0tL3ybeLqC5xbNVY0-8iA&redir_esc=y#v=onepage&q=bulletin FAO 39 Lysimeters&f=false

Agarwal, A., Prakash, O., & Bala, M. (2021). Effect of irrigation schedule on growth and seed yield of camelina (Camelina sativa L.) in Tarai region of central Himalaya. *Oil Crop Science,6* (1), 8-11. https://doi.org/10.1016/j.ocsci.2021.01.004

Ahmed, Z., Liu, J., Ahmad, E., Id, W., Yan, Y., & Qi, Z. (2020). *Differential physio-biochemical and yield responses of Camelina sativa L . under varying irrigation water regimes in semi-arid climatic conditions.PLoS ONE. Vol.15*(12), 1–18. https://doi.org/10.1371/journal.pone.0242441

Ahmed, Z., Waraich, E. A., Qi, Z., Shareef, Muhammad Waraich, E., Gui, D., Shabbir, N., & Iqbal, H. (2019). Physio-biochemical and yield responses of two contrasting Camelina sativa L . breeding lines under drought stress Physio-Biochemical and Yield Responses of Two Contrasting Camelina sativa Breeding Lines under Drought Stress.Intl. J. Agric. Biol,22(5). *International Journal of Agriculture & Biology,22* (5), 1186-1196. https://doi.org/10.17957/IJAB/15.1186

Aiken, R., Baltensperger, D., Krall, J., Pavlista, A., & Johnson, J. (2015). Planting methods affect emergence , flowering and yield of spring oilseed crops in the U . S . central High Plains. *Industrial Crops & Products,69* , 273-277. https://doi.org/10.1016/j.indcrop.2015.02.025

Akkaya, A. (1994). *Wheat Cultivation. Sütc, üimam University Textbooks Publication No.1, Kahramanmaras, (in Turkish).*

Amiri-Darban, N., Nourmohammadi, G., Hosein, A., Rad, S., Mohammad, S., Mirhadi, J., & Majidi, I. (2020). Industrial Crops & Products Potassium sulfate and ammonium sulfate a ff ect quality and quantity of camelina oil grown with di ff erent irrigation regimes. *Industrial Crops & Products,148* , 1-8. https://doi.org/10.1016/j.indcrop.2020.112308

Asseng, S., Ewert, F., Martre, P., Rötter, R. P., Lobell, D. B., Cammarano, D., Kimball, B. A., Ottman, M. J., Wall, G. W., White, J. W., Reynolds, M. P., Alderman, P. D., Prasad, P. V. V, Aggarwal, P. K., Anothai, J., Basso, B., Biernath, C., Challinor, A. J., De Sanctis, G., ... Zhu, Y. (2015). Rising temperatures reduce global wheat production. *Nature Climate Change,5* (2), 143-147. https://doi.org/10.1038/nclimate2470

Belayneh, H. D., Wehling, R. L., Cahoon, E., & Ciftci, O. N. (2015). supercritical carbon dioxide. *The Journal of Supercritical Fluids,104* , 153-159. https://doi.org/10.1016/j.supflu.2015.06.002

Berenguer, M., Vossen, P., Grattan, S., Connell, J., & Polito, V. (2006). Tree Irrigation Levels for Optimum Chemical and Sensory Properties of Olive Oil. *HortScience,41* , 427-432. https://doi.org/10.21273/HORTSCI.41.2.427

Berti, M., Gesch, R., Johnson, B., Ji, Y., Seames, W., & Aponte, A. (2015). Double- and relay-cropping of energy crops in the northern Great Plains, USA. *Industrial Crops and Products,75* , 26-34. https://doi.org/https://doi.org/10.1016/j.indcrop.2015.05.012

Bilalis, D., Roussis, I., Fuentes, F., Kakabouki, I., & Travlos, I. (2017). Organic Agriculture and Innovative Crops under Mediterranean Conditions. *Notulae Botanicae HortiCluj--Napoca,45* (2), 323-331. https://doi.org/10.15835/nbha45210867

Blum, A. (2009). Effective use of water (EUW) and not water-use efficiency (WUE) is the target of crop yield improvement under drought stress. *Field Crops Research,112* , 119-123. https://doi.org/10.1016/j.fcr.2009.03.009

Capra, A., Consoli, S., & Scicolone, B. (2008). No Title. In A. Alonso & H. Iglesias (Eds.), *Agricultural Irrigation Research Progress*. Nova Scienc Nova Science Publishers, Inc.e.

Carton, N., Carlsson, G., Jensen, E., & Corre-Hellou, G. (2020). *Intercropping of lupin for weed control and complementary grain production in organic farming.* https://doi.org/10.5281/zenodo.3741486

Chai, Q., Gan, Y., Zhao, C., Xu, H., Waskom, R., Niu, Y., & Siddique, K. (2016). Regulated deficit irrigation for crop production under drought stress. A review. *Agronomy for Sustainable Development,36* . https://doi.org/10.1007/s13593-015-0338-6

Chalmers, D., Mitchell, P., & Heek, L. (1981). Control of Peach Tree Growth and Productivity by Regulated Water Supply, Tree Density, and Summer Pruning1. *Journal of the American Society for Horticultural Science,106* , 307-312. https://doi.org/10.21273/JASHS.106.3.307

Cheng, L., Wang, Y., He, Q., Li, H., Zhang, X., & Zhang, F. (2016). Comparative proteomics illustrates the complexity of drought resistance mechanisms in two wheat (Triticum aestivum L .) cultivars under dehydration and rehydration. *BMC Plant Biology,16* , 1-23. https://doi.org/10.1186/s12870-016-0871-8

Cheng, X., Zhang, Q., Shi, C., & Yan, X. (2017). Model for the prediction of grain density and pressure distribution in hopper-bottom silos. *Biosystems Engineering,163* , 159-166. https://doi.org/https://doi.org/10.1016/j.biosystemseng.2017.09.006

Chennafi, H., Aidaoui, A., Bouzezour, H., & Sasi, A. (2006). Yield response of Durum Wheat (Triticum durum Desf.) Cultivar Waha to deficit Irrigation Under semi-Arid Growth Conditions. *Asian Journal of Plant Sciences,5* (5), 854-860. https://doi.org/10.3923/ajps.2006.854.860b

Colburn, M. W., Evans, J. L., & Ramage, C. H. (1968). Apparent and True Digestibility of Forage Nutrients by Ruminant Animals. *Journal of Dairy Science,51* (9), 1450-1457. https://doi.org/10.3168/jds.S0022-0302(68)87211-X

Czarnik, M., Jarecki, W., & Bobrecka-jamro, D. (2018). Reaction of winter varieties of false flax (Camelina sativa (L .) Crantz) to the varied sowing time. *Journal of Central European Agriculture, 19*(3), 571–586.

Daran, R., Suat, I., Richard, F., Tim, S., Koffi, D., Glen, S., Aaron, B., Nicholas, W., Dennis, F., Marty, S., Brian, W., & Simon, V. D. (2016). Economic Return versus Crop Water Productivity of Maize for Various Nitrogen Rates under Full Irrigation, Limited Irrigation, and Rainfed Settings in South Central Nebraska. *Journal of Irrigation and Drainage Engineering,142* (6), 4016017. https://doi.org/10.1061/(ASCE)IR.1943-4774.0001023

Delmas, T., Han, S., Kim, J., Bibette, J., Vauquelin, R., Sas, C., & C, H. B. (2013). *Skin cell targeting with self-assembled ligand addressed nanoemulsion droplets*. 310–318.

https://doi.org/10.1111/ics.12044

Dobrzyńska, M. A., & Przysławski, J. (2020). The effect of camelina oil (α -linolenic acid) and canola oil (oleic acid) on lipid profile , blood pressure , and anthropometric parameters in postmenopausal women. *Archives of Medical Science,17* (06), 1566-1574. https://doi.org/10.5114/aoms.2020.94033

Donald, C., & Hamblin, J. (1976). The biological yield and the harvest index of cereals as agronomic and plant selection criteria. *Advances in Agronomy,28* , 361-405. https://doi.org/10.1016/S0065-2113(08)60559-3

Evrard, J. (2005). Les tourteaux d ' oléagineux , source de protéines en alimentation animale. *Oilseeds and Fats, Crops and Lipids,12* (03), 224-227. http://www.lrrd.org/lrrd22/8/chab22153.htm

Fabeiro, C., Olalla, F., López-Urrea, R., & Domínguez, A. (2003). Production and quality of the sugar beet (Beta vulgaris L.) cultivated under controlled deficit irrigation conditions in a semi-arid climate. *Agricultural Water Management,62* , 215-227. https://doi.org/10.1016/S0378-3774(03)00097-0

FAO. (2011). *Producing more with less* (FAO (ed.)). https://www.fao.org/3/i2215f/i2215f.pdf

FAO. (2015). *Algeria*. 2019.

FAO. (2022). *World Food and Agriculture - Statistical Yearbook 2022. Rome* (FAO (ed.)). FAO. https://doi.org/https://doi.org/10.4060/cc2211en

Fernandes-Silva, A., Oliveira, M., Paço, T., & Ferreira, M. (2018). Deficit Irrigation in Mediterranean Fruit Trees and Grapevines: Water Stress Indicators and Crop Responses. In *Irrigation in Agroecosystems* (pp. 52-85). https://doi.org/DOI: 10.5772/intechopen.80365

French, U. N., Hunsaker, D., Thorp, K., & Clarke, T. (2009). Evapotranspiration over a camelina crop at Maricopa, Arizona. *Industrial Crops and Products,29* (2-3), 289-300. https://doi.org/10.1016/j.indcrop.2008.06.001

Gaba, S., Lescourret, F., Boudsocq, S., & Enjalbert, J. (2015). Multiple cropping systems as drivers for providing multiple ecosystem services: from concepts to design. *Agronomy for Sustainable Development,35* , 607-623. https://doi.org/10.1007/s13593-014-0272-z

Gan, Y., Siddique, K., Turner, N., X-G, L., Niu, J.-Y., Yang, C., Liu, L., & Chai, Q. (2013). Ridge-furrow mulching systems-an innovative technique for boosting crop productivity in semiarid rain-fed environments. *Adv Agron,118* , 429-476. https://doi.org/https://doi.org/10.1007/s11104-010-0312-7

Geerts, S., & Raes, D. (2009). Deficit irrigation as an on-farm strategy to maximize crop water productivity in dry areas. *Agricultural Water Management,96* (9), 1275-1284. https://doi.org/10.1016/j.agwat.2009.04.009

Girona, J., Gelly, M., Mata, M., Arbonès, A., Pascual, M., & Marsal, J. (2005). Peach tree response to single an.combined deficit irrigation regimes in deep soils. *Agricultural Water Management,72* , 97-108. https://doi.org/10.1016/j.agwat.2004.09.011

Gollner, G., Gabler, C., Grausgruber-Gröger, S., Friedel, J., Grausgruber, H., & Freyer, B. (2010). Grain legumes in mixed stands with false flax (camelina sativa) in organic farming under pannonian site conditions. *Journal Fur Kulturpflanzen, 62*, 402–408.

Grassini, P., Yang, H., Irmak, S., Thorburn, J., Burr, C., & Cassman, K. G. (2011). High-yield irrigated maize in the Western U.S. Corn Belt: II. Irrigation management and crop water productivity. *Field Crops Research,120* (1), 133-141. https://doi.org/https://doi.org/10.1016/j.fcr.2010.09.013

Gregory, P. J., Ingram, J. S., & Brklacich, M. (2005). Climate change and food security. *Philosophical Transactions of the Royal Society of London. Series B, Biological Sciences,360* (1463), 2139-2148. https://doi.org/https://doi.org/10.1098/rstb.2005.1745

Groeneveld, J. H., & Klein, A.-M. (2014). Pollination of two oil-producing plant species: Camelina (Camelina sativa L. Crantz) and pennycress (Thlaspi arvense L.) double-cropping in Germany. *GCB Bioenergy,6* (3), 242-251. https://doi.org/https://doi.org/10.1111/gcbb.12122

Guedes, W., Nobre, R., Soares, L., Lima, G., Gheyi, H., Fernandes, P., Ferreira, A., Rodrigues da Silva, A., Azevedo, C., Silva, D., & Medeiros, J. (2023). Irrigation Strategies with Controlled Water Deficit in Two Production Cycles of Cotton. *Plants,12* , 2892. https://doi.org/10.3390/plants12162892

Guendouz, A., Hannachi, A., & Benidir, M. (2022). Agro-biochemical Characterisation of Camelina sativa: A Review Agro-biochemical Characterisation of Camelina sativa : A Review. *Agricultural Reviews,43* (3), 278-287. https://doi.org/10.18805/ag.RF-230

Gülen, S. C. (2011). A simple parametric model for the analysis of cooled gas turbines. *Journal of Engineering for Gas Turbines and Power,133* (1), 11801. https://doi.org/10.1115/1.4001829

Hamdy, A., Sardo, V., & Ghanem, K. (2005). Saline water in supplemental irrigation of wheat and barley under rain fed agriculture. *Agricultural Water Management, 78*, 122–127.

Hélène, D. C. (2020). *Functional roles of lipids in make-up products.*

Hergert, G. W., Margheim, J. F., Pavlista, A. D., Martin, D. L., Isbell, T. A., & Supalla, R. J. (2016). Irrigation response and water productivity of deficit to fully irrigated spring camelina. *Agricultural Water Management,177* , 46-53. https://doi.org/10.1016/j.agwat.2016.06.009

Hixson, S. M., Parrish, C. C., & Anderson, D. M. (2014). Full substitution of fish oil with camelina (Camelina sativa) oil, with partial substitution of fish meal with camelina meal, in diets for farmed Atlantic salmon (Salmo salar) and its effect on tissue lipids and sensory quality. *Food Chemistry,157* , 51-61. https://doi.org/https://doi.org/10.1016/j.foodchem.2014.02.026

Howell, T., Lamm, F., & Tolk, J. (2010). Deficit Irrigation Management: Concepts and Implementation. *5th National Decennial Irrigation Conference Proceedings, 5-8 December 2010, Phoenix Convention Center, Phoenix, Arizona USA.* https://doi.org/https://doi.org/10.13031/2013.35864

Hrastar, R., MARINKA G, P., NIVES, O., & IZTOK JOˇZE, K. (2009). Fatty Acid and Stable Carbon Isotope Characterization of Camelina sativa Oil: Implications. *Journal of Agricultural and Food Chemistry, 57*(2), 579–585.

Hunsaker, D., French, A., & Elshikha, D. E. M. (2011). Water use , crop coefficients , and irrigation management criteria for camelina production in arid regions. *Irrigation Science,29* (1), 27-43. https://doi.org/10.1007/s00271-010-0213-9

Hunsaker, D. J., French, A. N., Clarke, T. R., & El-Shikha, D. M. (2011). Water use, crop coefficients, and irrigation management criteria for camelina production in arid regions. *Irrigation Science,29* (1), 27-43. https://doi.org/10.1007/s00271-010-0213-9

Hunsaker, D. J., French, A. N., & Thorp, K. R. (2013). Camelina water use and seed yield response to irrigation scheduling in an arid environment. *Irrigation Science,31* , 911-929. https://doi.org/10.1007/s00271-012-0368-7

Intrigliolo, D. S. (2020). DEFICIT IRRIGATION STRATEGIES AND KEY RESULTS IN REAL CASE STUDIES. *FAO.* https://www.fao.org/3/cb5065fr/cb5065fr.pdf

Jankowski, K. J. (2019). Camelina: Yield and quality response to nitrogen and sulfur fertilization in Poland. *Industrial Crops & Products,141* (February), 1-10. https://doi.org/10.1016/j.indcrop.2019.111776

Johnson, N., C, R., & J, E. (2001). Managing water for people and nature. *Science, 292,* 1071–1072.

Juodka, R., Nainienė, R., Juškienė, V., Juška, R., Leikus, R., Kadžienė, G., & Stankevičienė, D. (2022). Camelina (Camelina sativa (L.) Crantz) as Feedstuffs in Meat Type Poultry Diet: A Source of Protein and n-3 Fatty Acids. *Animals,12* (3). https://doi.org/10.3390/ani12030295

Katarzyna, R., Edyta, S., Małgorzata, W., & Magdalena, R. 'nska (2018). Bioactive Compounds , Nutritional Quality and Oxidative Stability of Cold-Pressed Camelina (Camelina sativa L .) Oils. *Applied Sciences,8* (12), 1-17. https://doi.org/10.3390/app8122606

Kemp, W. H. (2006). *Biodiesel basics and beyond: a comprehensive guide to production and use for the home and farm*. Aztext press. https://cir.nii.ac.jp/crid/1130000796780723584.bib?lang=en

Kessai, A. K., Debabeche, M., & Pitacco, A. (2018). *application of deficit irrigation strategy for barley and wheat in the region of Biskra.*

Khemmouli, A., Houria, C., Abderrahmane, H., Samir, M., Yamina, H., Rekik, F., Asma, M., Cheli, R., & Ghougali, F. (2024). Effects of deficit irrigation on the composition of by-products of camelina sativa: towards sustainable water use in semi-arid agriculture. *Natural Resources and Sustainable Development Volume,14* (1), 165-182. https://doi.org/10.31924/nrsd.v14i1.158

Kirkhus, B., Lundon, A. R., Haugen, J., Vogt, G., Borge, G. I. A., & Henriksen, B. I. F. (2013). Effects of Environmental Factors on Edible Oil Quality of Organically Grown Camelina sativa. *Journal of Agricultural and Food Chemistry*, *61*, 3179–3185.

Kisekka, I., P. Aguilar, J., H. Rogers, D., Holman, J., M. O'Brien, D., & Klocke, N. (2016). Assessing Deficit Irrigation Strategies for Corn Using Simulation. *Transactions of the ASABE,59* (1), 303-317. https://doi.org/https://doi.org/10.13031/trans.59.11206

Krohn, B. J., & Fripp, M. (2012). A life cycle assessment of biodiesel derived from the '"niche filling"' energy crop camelina in the USA. *Applied Energy,92* , 92-98. https://doi.org/10.1016/j.apenergy.2011.10.025

Krzyżaniak, M., Stolarski, M. J., Tworkowski, J., Puttick, D., & Eynck, C. (2019). Industrial Crops & Products Yield and seed composition of 10 spring camelina genotypes cultivated in the temperate climate of Central Europe. *Industrial Crops & Products,138* , 1-9. https://doi.org/10.1016/j.indcrop.2019.06.006

Kumar, A., Sharma, K., & Yadav, A. (2010). Enhancing yield and water productivity of wheat (Triticum aestivum) through furrow igated raised bed system in the indo Gangetic Plains of India. *Indian J Agric Sci, 80*(3), 198–202.

Leclère, D., Havlík, P., Fuss, S., Schmid, E., Mosnier, A., Walsh, B., Valin, H., Herrero, M., Khabarov, N., & Obersteiner, M. (2014). Climate change induced transformations of agricultural systems: Insights from a global model. *Environmental Research Letters,9* . https://doi.org/10.1088/1748-9326/9/12/124018

Leclère, M., Jeu, M., Butier, A., Chatain, C., & Loyce, C. (2019). Controlling weeds in camelina with innovative herbicide-free crop management routes across various environments. *Industrial Crops & Products Journal,140* (July), 111605. https://doi.org/10.1016/j.indcrop.2019.111605

Lopez-Gunn, E., Zorrilla, P., Prieto, F., & Llamas, M. R. (2012). Water efficiency in Spanish agriculture. *Agricultural Water Management,108* , 83-95. https://doi.org/10.1016/j.agwat.2012.01.005

Lopez, J. R., Winter, J. M., Elliott, J., Ruane, A. C., & Porter, C. (2017). Agricultural and Forest Meteorology Integrating growth stage de fi cit irrigation into a process based crop model. *Agricultural and Forest Meteorology,243* (December 2016), 84-92. https://doi.org/10.1016/j.agrformet.2017.05.001

Macedo, F. G. De, Bresolin, J. D., Santos, E. F., Furlan, F., Lopes, W. T., Polacco, J. C., Lavres, J., & Sperotto, R. A. (2016). Nickel Availability in Soil as Influenced by Liming and Its Role in Soybean Nitrogen Metabolism. *Frontiers in Plant Science,7* (September), 1-12. https://doi.org/10.3389/fpls.2016.01358

Martinelli, T., & Galasso, I. (2011). Phenological growth stages of Camelina sativa according to the extended BBCH scale. *Annals of Applied Biology,158* (1), 87-94. https://doi.org/10.1111/j.1744-7348.2010.00444.x

Masella, P., Martinelli, T., & Galasso, I. (2014). Agronomic evaluation and phenotypic plasticity of *Camelina sativa* growing in Lombardia, Italy. *Crop and Pasture Science,65* (5), 453-460. https://doi.org/10.1071/CP14025

McDermid, S., Nocco, M., Lawston-Parker, P., Keune, J., Pokhrel, Y., Jain, M., Jägermeyr, J., Brocca, L., Massari, C., Jones, A. D., Vahmani, P., Thiery, W., Yao, Y., Bell, A., Chen, L., Dorigo, W., Hanasaki, N., Jasechko, S., Lo, M.-H., ... Yokohata, T. (2023). Irrigation in the Earth system. *Nature Reviews Earth & Environment,4* (7), 435-453. https://doi.org/10.1038/s43017-023-00438-5

McMaster, G. S., & Wilhelm, W. W. (1997). Growing degree-days: one equation, two interpretations. *Agricultural and Forest Meteorology,87* , 291-300. https://doi.org/10.1016/S0168-1923(97)00027-0

Mesbah, E. A. E. (2009). Effect of irrigation regimes and foliar spraying of potassium on yield, yield component and water use efficiency in wheat (Triticum aestivum L.) in sandy soils. *World Journal of Agricultural Sciences,5* (6), 662-669. https://www.idosi.org/wjas/wjas5(6)/1.pdf

Meyer, N., Jacques-Eric, B., Julie, C., & Eric, J. (2019). Cover crops reduce water drainage in temperate climates: A meta-analysis. *Agronomy for Sustainable Development,39* , 1-11. https://doi.org/https://doi.org/10.1007/s13593-018-0546-y

Mortlock, M. Y., & Vanderlip, R. L. (1989). Germination and establishment of pearl millet and sorghum of different seed qualities under controlled high-temperature environments. *Field Crops Research,22* (3), 195-209. https://doi.org/https://doi.org/10.1016/0378-4290(89)90092-0

Mukherjee, S., Dash, P., Das, D., & Das, S. (2023). Growth, Yield and Water Productivity of Tomato as Influenced by Deficit Irrigation Water Management. *Environmental Processes,10* . https://doi.org/10.1007/s40710-023-00624-z

Musick, J., Jones, O., Stewart, B., & Dusek, D. (1994). Water-Yield Relationships for Irrigated and Dryland Wheat in the U.S. Southern Plains. *Agronomy Journal,86* , 980-986. https://doi.org/10.2134/agronj1994.00021962008600060010x

Nadhira, A., & Ouahiba, B. (2018). Algeria towards Sustainable Agricultural Development. *Bulletin de La Société de Géographie d'Egypte,91* (1), 49-66. https://doi.org/10.21608/bsge.2018.90306

Neupane, D., Solomon, J. K. Q., McLennon, E., Davison, J., & Lawry, T. (2020). Camelina production parameters response to different irrigation regimes. *Industrial Crops and Products,148* , 1-9. https://doi.org/10.1016/j.indcrop.2020.112286

Nithyapriya, S., Lalitha, S., Sayyed, R. Z., Reddy, M. S., Dailin, D. J., El Enshasy, H. A., Luh Suriani, N., & Herlambang, S. (2021). Production, Purification, and Characterization of Bacillibactin Siderophore of Bacillus subtilis and Its Application for Improvement in Plant Growth and Oil Content in Sesame. *Sustainability,13* (10). https://doi.org/10.3390/su13105394

Noreldin, T., Ouda, S., Mounzer, O., & Abdelhamid, M. (2015). CropSyst model for wheat under deficit irrigation using sprinkler and drip irrigation in sandy soil. *Journal of Water and Land Development,26* . https://doi.org/10.1515/jwld-2015-0016

Obour, A. K., Obeng, E., Mohammed, Y. A., Ciampitti, I. A., Durrett, T. P., Aznar-Moreno, J. A., & Chen, C. (2017). Camelina Seed Yield and Fatty Acids as Influenced by Genotype and Environment. *Agronomy Journal,109* (3), 947-956. https://doi.org/https://doi.org/10.2134/agronj2016.05.0256

Ouda, S., Noreldin, T., Alarcón, J. J., Ragab, R., Caruso, G., Sekara, A., & Abdelhamid, M. T. (2021). Response of spring wheat (Triticum aestivum) to deficit irrigation management under the semi-arid environment of egypt: Field and modeling study. *Agriculture (Switzerland),11* (2), 1-13. https://doi.org/10.3390/agriculture11020090

Paula, E. M., da Silva, L. G., Brandao, V. L. N., Dai, X., & Faciola, A. P. (2019). Feeding Canola, Camelina, and Carinata Meals to Ruminants. *Animals,9* (10), 1-19. https://doi.org/10.3390/ani9100704

Paulsen, H. (2011). Improving green-house gas balances of organic farms by the use of straight vegetable oil from mixed cropping as farm own fuel and its competition to food production. *Landbauforschung Volkenrode*, *61*, 209–216.

Pavlista, A. D., Hergert, G. W., Margheim, J. M., & Isbell, T. A. (2016). Growth of spring camelina (Camelina sativa) under deficit irrigation in Western Nebraska. *Industrial*

Crops & Products,83 , 118-123. https://doi.org/10.1016/j.indcrop.2015.12.017

Pavlista, A. D., Isbell, T. A., Baltensperger, D. D., & Hergert, G. W. (2011). Planting date and development of spring-seeded irrigated canola, brown mustard and camelina. *Industrial Crops and Products,33* (2), 451-456. https://doi.org/10.1016/J.INDCROP.2010.10.029

Payero, J. O., Tarkalson, D. D., Irmak, S., Davison, D., & Peterson, J. L. (2008). Effect of irrigation amounts applied with subsurface drip irrigation on corn evapotranspiration, yield, water use efficiency, and dry matter production in a semiarid climate . *Agricultural Water Management,95* (08), 895-908. https://doi.org/10.1016/j.agwat.2008.02.015

Petit, S., Cordeau, S., Chauvel, B., Bohan, D., Guillemin, J.-P., & Steinberg, C. (2018). Biodiversity-based options for arable weed management. A review. *Agronomy for Sustainable Development,38* (5), 48. https://doi.org/10.1007/s13593-018-0525-3

Ramadan, A., El Habbasha, E. S., Taha, M., & Refaie, K. (2014). Effect of Irrigation Water Requirements and Fertigation Levels on Growth, Yield and Water Use Efficiency in Wheat. *Middle East Journal of Scientific Research,16* , 441-450. https://doi.org/10.5829/idosi.mejsr.2013.16.04.11733

Rastoin, J., & Benabderrazik, E. H. (2014). *et oléoprotéagineux au Maghreb Céréales*. 1–32.

Raza, M. A. S., Shahid, A. M., Ijaz, M., Khan, I. H., Saleem, & Ahmad, S. (2015). Studies on canola (Brassica napus L.) and camelina (Camelina sativa L.) under different irrigation levels. *Journal of Agricultural and Biological Science,10* (4), 130-138. https://doi.org/i

Rekaby, S. A., Eissa, M. A., Hegab, S. A., & Ragheb, H. M. (2016). Effect of nitrogen fertilization rates on wheat grown under drip irrigation. *Assiut Journal of Agricultural Sci Encesences,47* (3), 104-119. https://doi.org/doi.org / 10.21608 / ajas.2016908.

Richard, S., Morency, M. J., Drevet, C., Jouanin, L., & Séguin, A. (2000) Isolation and characterization of a dehydrin gene from white spruce induced upon wounding, drought and cold stresses. *Plant Molecular Biology,43* , 1-10. https://doi.org/10.1023/A:1006453811911

Righini, D., Zanetti, F., Martínez-Force, E., Mandrioli, M., Gallina, T., & Monti, A. (2019). Shifting sowing of camelina from spring to autumn enhances the oil quality for bio-based applications in response to temperature and seed carbon stock. *Industrial Crops & Products,137* , 66-73. https://doi.org/10.1016/j.indcrop.2019.05.009

Roxana, B. (2024). *NASA Analysis Confirms 2023 as Warmest Year on Record*. 12 Jan 2024. https://www.nasa.gov/news-release/nasa-analysis-confirms-2023-as-warmest-year-on-record/

Royo-Esnal, A., & Valencia-Gredilla, F. (2018). Camelina as a Rotation Crop for Weed Control in Organic Farming in a Semiarid Mediterranean Climate. *Agriculture,156* , 1-12. https://doi.org/10.3390/agriculture8100156

SAUCKE, H., & ACKERMANN, K. (2006). Weed suppression in mixed cropped grain peas and false flax (Camelina sativa). *Weed Research,46* (6), 453-461. https://doi.org/https://doi.org/10.1111/j.1365-3180.2006.00530.x

Shang, H. Y. (2015). Emergy calculation and analysis of food crops in Zhangye city. *Advance Journal of Food Science and Technology,7*(12), 921–925.

https://doi.org/10.19026/ajfst.7.2534

Sindelar, A. J., Schmer, M. R., & Gesch, R. W. (2017). Winter oilseed production for biofuel in the US Corn Belt: opportunities and limitations. *GCB Bioenergy,9* , 508-524. https://doi.org/10.1111/gcbb.12297

Singh, Y., Cullere, M., Tůmová, E., & Zotte, A. D. (2023). Camelina sativa as a sustainable and feasible feedstuff for broiler poultry species: A review. *Czech Journal of Animal Science,68* (7), 277-295. https://doi.org/10.17221/29/2023-CJAS

Sintim, H. Y., Zheljazkov, V. D., Obour, A. K., Garcia, A., & Foulke, T. K. (2015). Influence of nitrogen and sulfur application on camelina performance under dryland conditions. *Industrial Crops & Products,70* , 253-259. https://doi.org/10.1016/j.indcrop.2015.03.062

Sofo, A., Palese, A., Casacchia, T., Dichio, B., Xiloyannis, C, N. Y., & 34-1-6. (2012). Sustainable fruit production in Mediterranean orchards subjected to drought stress. In *Mediterranean orchards subjected to drought stress. In: Ahmad P, Prasad MNV (eds) Abiotic stress responses in plants metabolism, productivity and sustainability* (pp. 105-129.). Springer. https://doi.org/https://doi.org/10.1007/978-1-4614-0634-1-6

Stanhill, G. (1986). Water use efficiency. Adv Agron. *Adv Agron, 39*, 53–85.

Sterligov, V. V. (2020). Prediction of carbon dioxide emissions from fuel combustion. *IOP Conference Series: Materials Science and Engineering,866* (1), 12028. https://doi.org/10.1088/1757-899X/866/1/012028

Struik, P. C., & Kuyper, T.. (2017). Sustainable intensification in agriculture: the richer shade of green. A review. *Agronomy for Sustainable Development,37* , 1-15. https://doi.org/10.1007/s13593-017-0445-7

Sydor, M., Kurasiak-Popowska, D., & Stuper-Szablewska, K. (2022). Industrial Crops & Products Camelina sativa. Status quo and future perspectives. *Industrial Crops & Products Journal,187* , 1-12. https://doi.org/10.1016/j.indcrop.2022.115531

Taha, A., Ouda, S., & Zohry, A. E.-H. (2015). *High Water Consuming Crops Under Control: Case of Sugarcane Crop* (pp. 85-97). https://doi.org/10.1007/978-3-319-21771-0_6

Tari, A. F. (2016). The effects of different deficit irrigation strategies on yield, quality, and water-use efficiencies of wheat under semi-arid conditions. *Agricultural Water Management,167* , 1-10. https://doi.org/10.1016/j.agwat.2015.12.023

UNEPMAP - UN. (n.d.). *Climate change in the Mediterranean.* https://www.unep.org/unepmap/resources/factsheets/climate-change

Uremis, I., Arslan, M., Sangun, M. K., & Uygur, V. (2009). Allelopathic potential of rapeseed cultivars on germination and seedling growth of weeds Allelopathic Potential of Rapeseed Cultivars on Germination and Seedling Growth of Weeds. *Asian Journal of Chemistry -, 21*(3), 2170–2184.

Van Reeuwijk, L. P. (1995). *Soil analysis procedures, Technical Document 9, 5th edition, International Soil Reference and Information Centre, FAO.* https://www.isric.org/sites/default/files/ISRIC_TechPap09.pdf

Vollmann, J., & Eynck, C. (2015). Camelina as a sustainable oilseed crop: Contributions of plant breeding and genetic engineering. *Biotechnology Journal,10*, 525–535.

https://doi.org/10.1002/biot.201400200

Wysocki, D. J., Chastain, T. G., Schillinger, W. F., Guy, S. O., & Karow, R. S. (2013). Camelina: Seed yield response to applied nitrogen and sulfur. *Field Crops Research,145* , 60-66. https://doi.org/https://doi.org/10.1016/j.fcr.2013.02.009

Zajac, M., Bozena, K., Wioletta, S., & Renata, K. (2020). Inclusion of Camelina, Flax, and Sunflower Seeds in the Diets for Broiler Chickens: Apparent Digestibility of Nutrients, Growth Performance, Health Status, and Carcass and Meat Quality Traits. *Animals,10* (2), 1-17. https://doi.org/10.3390/ani10020321

Załuski, D., Tworkowski, J., Krzyzaniak, M., Stolarski, M. J., & Kwiatkowski, J. (2020). The Characterization of 10 Spring Camelina Genotypes Grown in Environmental Conditions in. *Agronomy,10* (1), 1-13. https://doi.org/https://doi.org/10.3390/agronomy10010064

Zanetti, F., Alberghini, B., Jérômela, A. M., Grahovac, N., Rajkovi, D., Kiprovski, B., & Monti, A. (2021). *Camelina , an ancient oil crop actively contributing to the rural renaissance in Europe . A review 2 Morphological description and main The interest in camelina is attested by at least 100 articles per year on this species.*. 1–18.

Zanetti, F., Alberghini, B., Marjanović Jeromela, A., Graovac, N., Rajković, D., Kiprovska, B., & Monti, A. (2021). Camelina, an ancient oilseed crop actively contributing to the rural renaissance in Europe. A review. *Agronomy for Sustainable Development,41* , 1-18. https://doi.org/10.1007/s13593-020-00663-y

Zanetti, F., Eynck, C., Christou, M., Krzyżaniak, M., Righini, D., Alexopoulou, E., Stolarski, M. J., Van Loo, E. N., Puttick, D., & Monti, A. (2017). Agronomic performance and seed quality attributes of Camelina (Camelina sativa L. crantz) in multi-environment trials across Europe and Canada. *Industrial Crops and Products,107* , 602-608. https://doi.org/10.1016/j.indcrop.2017.06.022

Zhang, B., Li, F., Huang, G., Cheng, Z., & Zhang, Y. (2006). Yield of spring wheat improved by regulated deficit irrigation in an arid zone. *Agricultural Water Management,79* , 28-42. https://doi.org/10.1016/j.agwat.2005.02.007

Zhang, C., & Auer, C. (2020). Industrial Crops & Products Hybridization between Camelina sativa (L .) Crantz and common Brassica weeds. *Industrial Crops & Products,147* (September 2019), 112240. https://doi.org/10.1016/j.indcrop.2020.112240

Zubr, J. (1997). *Oil-seed crop: Camelina sativa. 6690*(96).

Zubr, J. (2010). Carbohydrates , vitamins and minerals of Camelina sativa seed. *Nutrition & Food Science,40* (5), 523-531. https://doi.org/10.1108/00346651011077036

FSC
www.fsc.org
MIX
Papier | Fördert
gute Waldnutzung
FSC® C083411

Kolibri 360 GmbH
Ferdinand-Jühlke-Straße 7
99095 Erfurt, Deutschland
produktsicherheit@kolibri360.de

Druck:
CPI Druckdienstleistungen GmbH
im Auftrag der
Kolibri 360 GmbH
Ferdinand-Jühlke-Str. 7
99095 Erfurt